Industrial Gas Separations

Industrial Gas Separations

Thaddeus E. Whyte, Jr., EDITOR
Catalytica Associates, Inc.

Carmen M. Yon, EDITOR
Union Carbide Corporation

Earl H. Wagener, EDITOR
Dow Chemical Company

Based on a symposium sponsored
by the ACS Division of Industrial
and Engineering Chemistry,
Washington, D.C.,
June 14–15, 1982

ACS SYMPOSIUM SERIES **223**

AMERICAN CHEMICAL SOCIETY
WASHINGTON, D. C.   1983

Library of Congress Cataloging in Publication Data

Industrial gas separations.
    (ACS symposium series, ISSN 0097–6156; 223)

"Based on a symposium sponsored by the ACS
Division of Industrial and Engineering Chemistry,
Washington, D.C., June 14–15, 1982."

Includes bibliographies and index.

1. Gases—Separation—Congresses. I. Whyte,
Thaddeus E., 1937–   . II. Yon, Carmen M.,
1939–   . III. Wagener, Earl H. IV. American
Chemical Society. Division of Industrial and Engineering Chemistry. V. Series.

TP242.I65   1983   660.2842   83.6440
ISBN 0–8412–0780–1   1983

ACS Symposium Series

M. Joan Comstock, *Series Editor*

FOREWORD

The ACS Symposium Series was founded in 1974 to provide a medium for publishing symposia quickly in book form. The format of the Series parallels that of the continuing Advances in Chemistry Series except that in order to save time the papers are not typeset but are reproduced as they are submitted by the authors in camera-ready form. Papers are reviewed under the supervision of the Editors with the assistance of the Series Advisory Board and are selected to maintain the integrity of the symposia; however, verbatim reproductions of previously published papers are not accepted. Both reviews and reports of research are acceptable since symposia may embrace both types of presentation.

CONTENTS

Preface ix

1. Helium Recovery Using Semipermeable Membranes 1
   F. E. Martin, T. S. Snyder, and E. J. Lahoda

2. A Low-Temperature Energy-Efficient Acid Gas Removal Process . . . 27
   R. E. Hise, L. G. Massey, R. J .Adler, C. B. Brosilow,
   N. C. Gardner, L. Auyang, W. R. Brown, W. J. Cook, and
   Y. C. Liu

3. Implications of the Dual-Mode Sorption and Transport Models
   for Mixed Gas Permeation 47
   R. T. Chern, W. J. Koros, E. S. Sanders, S. H. Chen, and
   H. B. Hopfenberg

4. Standard Reference Materials for Gas Transmission Measurements . 75
   John D. Barnes

5. Gas Transport and Cooperative Main-Chain Motions in Glassy
   Polymers 89
   Daniel Raucher and Michael D. Sefcik

6. Sorption and Transport in Glassy Polymers: Gas–Polymer–Matrix
   Model 111
   Daniel Raucher and Michael D. Sefcik

7. Membrane Gas Separations for Chemical Processes and Energy
   Applications 125
   William J. Schell and C. Douglas Houston

8. Gas-Adsorption Processes: State of the Art 145
   George E. Keller, II

9. Nonisothermal Gas Sorption Kinetics 171
   Shivaji Sircar and Ravi Kumar

10. Recovery and Purification of Light Gases by Pressure Swing
    Adsorption 195
    Hsing C. Cheng and Frank B. Hill

11. Methane/Nitrogen Gas Separation over the Zeolite Clinoptilolite
    by the Selective Adsorption of Nitrogen 213
    T. C. Frankiewicz and R. G. Donnelly

12. Separation of Methane from Hydrogen and Carbon Monoxide by an
    Absorption/Stripping Process 235
    Vi-Duong Dang

13. **High-Temperature Hydrogen Sulfide Removal Using a Regenerable Iron Oxide Sorbent** **255**
   S. S. Tamhankar and C. Y. Wen

Index **281**

PREFACE

Nᴇᴡ ꜱᴇᴘᴀʀᴀᴛɪᴏɴ ᴛᴇᴄʜɴɪǫᴜᴇꜱ, particularly for improving gas technologies, have increased dramatically in recent years. These advances have stemmed not only from the obvious needs to reduce process costs and environmental pollution, but also from the synthesis and identification of novel materials. Such materials include unique membranes and zeolites.

The use of membranes to separate gases commercially is a relatively new application. Both Du Pont and Union Carbide had ventures in the early 1970s for recovering H_2 and He from industrial processes, but these projects were never fully commercialized. Recently, however, a combination of improved economics and better technology has resulted in membrane products that signal a new era in the commercial use of membranes for large-scale gas separation.

Diffusion processes in polymers have been studied since the 1860s, and gas diffusion in polymers was studied intensively in the 1950s and 1960s by such notables as Weller, Steiner, Kammermeyer, Stannett, Michaels, Rogers, and Stern. Although these studies generated a large amount of permeability data for gas diffusion through polymers, industrial applications were not implemented until nearly 15 years later. Clearly, the use of membranes to separate gases commercially is not a simple task. Companies currently trying to commercialize such processes are well aware of the need to combine basic understanding with sophisticated fabrication technology.

Both the understanding of gas transport and the number of engineering ventures that are using membranes have been growing rapidly. This symposium explores gas-transport mechanisms and models and presents several industrial applications of gas membranes. Although agreement on the transfer mechanism has not yet been reached for either mixed- or single-gas permeation, the understanding is sufficient to develop both energy- and cost-effective purification and bulk separation processes for industrial use.

Adsorption separation—an important unit operation in the separation of industrial gases—is achieved by adsorbing one or more component(s) onto the active sites of a solid adsorbent. For cyclic processes, the solid is regenerated by shifting the equilibrium toward the gas phase by increasing the adsorbent temperature or by decreasing the partial pressure of the adsorbate in the gas. Some of the more important commercial combinations of adsorbents and regeneration techniques are discussed in a paper in this symposium. As the understanding of cyclic adsorption processes has developed, more and more of the assumptions used in designing systems have been reexamined and discarded for more rigorous treatments. Many more sophisticated models of the adsorption phenomenon are being postulated, such as that offered for pressure swing separation of light gases. In other separations, the molecular size dimensions of the microporous solids are used to control kinetic selectivity as well as or in place of adsorptive selectivity. For example, the natural zeolite clinoptilolite is used for the separation of methane and nitrogen. In some adsorptive processes, the bonding energy associated with adsorption is similar in magnitude to that of a chemical reaction. For such cases, the regeneration may be achieved more effectively by shifting a chemical reaction equilibrium. Such an approach has been used for an iron oxide sorbent. The papers presented cover the spectrum of applications, adsorbents, and types of cycles.

Absorption is another unit operation that has many parallels to adsorption. Although the sorption media is liquid rather than solid, many of the underlying principles are the same. Most commercial applications for purification and bulk separation involve weak physical bonding of absorbate to absorbent. Again, regeneration of sorbent is accomplished by increasing the liquid temperature and/or by decreasing absorbate partial pressure. This volume contains a discussion of a process for recovering methane from hydrogen and carbon monoxide using liquid propane—a separation that might also have been attempted with a suitable solid adsorbent. Another state-of-the-art approach—the removal of acid gases (sulfur compounds and carbon dioxide)—is detailed; in this process the contaminant carbon dioxide becomes both an active adsorbent and a component of an absorptive fluid.

THADDEUS E. WHYTE, JR.
Catalytica Associates, Inc.
Santa Clara, CA

CARMEN M. YON
Union Carbide Corporation
Tarrytown, NY

EARL H. WAGENER
Dow Chemical Company
Walnut Creek, CA

February 28, 1983

Helium Recovery Using Semipermeable Membranes

F. E. MARTIN
Westinghouse Oceanic Division, Annapolis, MD 21404

T. S. SNYDER and E. J. LAHODA
Westinghouse Research & Development Center, Pittsburgh, PA 15235

A series of pure gas and gas mixture tests were performed, using a unique closed-circuit experimental loop. Based on data from continuous loop operations, a preliminary system design and economic evaluation were performed to recover helium from deep sea diving gas applications. This program was performed under Navy Contract N60921-79-C-A037 for the Naval Surface Weapons Center G-52. Membrane permeability was determined as a function of driving pressure and feed concentration for a single membrane element. Based on this data, a hypothetical design of a system to meet naval specifications was performed only as a contractual requirement with the Navy and is not intended as a specific Westinghouse system. Westinghouse at this time plans no such system for market. Projections based on the experimental data for the hypothetical system show that the least pure gas considered for the design, 58 mole % helium, could be enriched to better than 99.99% helium in five permeator stages. This gas could be enriched, hypothetically, to a physiologically acceptable quality in 3 stages. Carbon dioxide concentration in the gas is the design limiting parameter. This is a very conservative design estimate. The conservatism is necessary due to the limited nature of the design data.

The U. S. Navy, in pursuit of its many deep diving programs, must currently deal with exponentially rising costs and numerous supply difficulties concerning its prime diluent for breathing gas-helium. The logistics problems of providing large quantities of high pressure gas to remote diving sites over extended dive intervals, regardless of weather conditions, are difficult to quantify. However, provision for an indefinite supply of diluent via an on-board device that recycles expended gas, with relatively minor penalty to operational/maintenance costs and deck space,

would appear to be desirable. The benefit both to program
contingency planning and to dive safety could be substantial, even
if the device were relegated to a backup or supplemental mode of
operation.

More amenable to quantitative and objective analysis is the
economic incentive for developing a recycle system. It was
predicted in 1974(1) that, for helium recovery at the well-head,
a cost growth average of 17.9% per year could be anticipated for
the decade succeeding that report. However, it is further
predicted that natural gas fields bearing ≤ 0.3% helium will be
depleted by 1990-95, at which time fields bearing ≤ 0.1% must be
used. This would produce a sharp price escalation to wholesale
prices five times then-current rates. Thereafter, projections
are entirely dependent on how, whether, and when helium conservation and reserve storage policies are implemented, and on
whether future air separation technology will enable competitive
extraction of helium from the atmosphere.

The Navy, which, by our conservative estimate,(2) is a
current user of some 2,800,000 std ft^3 of helium per year, would
benefit substantially from a shipboard system that could capture
90-95% of currently wasted helium and condition it to acceptable
purity for re-use. The definition of acceptable purity is as
shown in Table I, wherein the design goals and design requirements
are set forth. The design goal reflects helium purity as found in
bottle supplies meeting the referenced specification, Grade A,
whereas the design requirement is derived from an assortment of
analyses based on the practical, operational constraints on a
diver's diluent gas. As per note 2 of Table I, the design
requirement can allow as much as 1% O_2 and 1% N_2, and consequent
drop of He to 97.8%, without impairing or jeopardizing the user;
however, the design was to include sizing to the 99.995% design
goal level to obtain an estimate for maximum design capacity
needed.

Experimental System Description

Figure 1 is a schematic of the experimental loop constructed
by the Westinghouse Oceanic Division to evaluate semipermeable
gaseous membranes for helium recovery. The key system components
in Figure 1 are the permeator modules, compressor, aftercooler,
humidifier and the online gas chromatograph for sample analyses.
This test system was unique in that it allowed closed circuit
flow of multicomponent gas mixtures to the membrane for separation,
followed by immediate recombination of the permeate and residual
streams for recycle to the feed side. This arrangement permitted
a constant volumetric feed rate to the membrane for prolonged
periods without requiring prohibitive and costly gas inventory.
Further, the experimental configuration provided complete monitoring and control capabilities over such parameters as back-pressure,
split-stream flow rates, gas temperature, humidity and composition.

Table I. Helium Purity Goals and Requirements

Component	Design[1] Goal	Design[1] Requirement	Rationale for Requirements
Helium (minimum %)	99.995	99.8[2]	Federal Specification BB-H-11680, Type I, Grade B
Water Vapor (ppm)	10	200	$30^\circ F$ dew point at 1000 ft
Oxygen (ppm)	2	1%	pO_2 = 0.3 atm at 850 ft
Nitrogen (%)	0	1%	Results in same pN_2 at 850 ft as initially pressurizing chamber to 14 ft with air
Carbon Dioxide (ppm)	0	100	pCO_2 = 0.003 atm at 850 ft
Carbon Monoxide (ppm)	0	12	OSHA exposure limit
Gaseous Hydrocarbons measured as Methane (ppm)	1	1	Federal Specification BB-H-11686, Type I, Grade C
Oil (mg/liter)	0	0.001	OSHA exposure limit
All others (ppm)	37	37	Federal Specification BB-H-11680, Type I, Grade A

NOTES: (1) Maximum values by volume, unless otherwise indicated
(2) May be further reduced by the amount of oxygen and nitrogen present

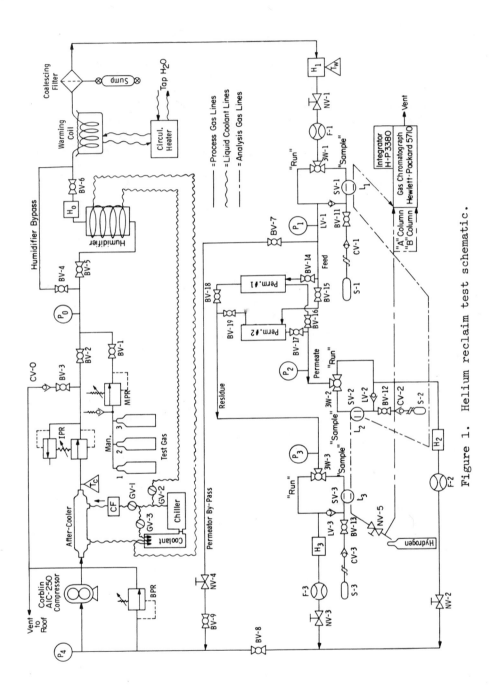

Figure 1. Helium reclaim test schematic.

PARAMETER	EQUIPMENT
F_1, F_2, F_3	Brooks Instruments' Model No. 5812 mass flow sensor and indicator/power supply model No. 5820, calibrated for 0-200 SLPM operation at 1000 psig. Helium, ±1% full-scale accuracy, <1% linearity from 1000 to 200 psig, 1-2% from 200⁺0 psig
L_1, L_2, L_3	Hewlett-Packard 5700A gas chromatograph, featuring an H-P 3380A data integrator, and utilizing Alltech Assoc. No. 5680PC packed columns of Carbosphere 60/80. Samples injected via Valco Instruments' Model No. 4V6-HPaX valve with 100 μ liter loop into a hydrogen carrier gas stream.
H_O, H_1, H_2, H_3	Panametrics' hygrometer/thermistor probe No. M221RT for use with No. 2830 sample cells and model no. 2100-151(F)-161(B) multi-channel hygrometer (-60 to +60°C dewpoint and -30 to +70°C temperature range, electrical accuracy 1% of input, readability 0.5% of full scale)
P_1, P_2, P_3	Matheson test gauge model 63-5613, 0-1400 psig, 0.25% accuracy
P_4, P_O	Matheson standard gauge model 63-3123, 0-2000 psig, 2-3% accuracy
CF	RCM Industries' liquid flowmeter, model 1/2 in.-7-1-R-2.
T_c	Established via Dunham-Bush chiller, model no. CCP-45Q per dwg. no. C2-C03828, providing coolant flow to compressor aftercooler; sensed by Panametrics M-2 thermister probe in no. 2830 cell.
T_w	Established via Precision Scientific circulating heater, Model 200, cat. no. 66613, serial no. 20 AK-5; sensed at H_1.

Additional equipment includes the manifold pressure regulator (MPR), Matheson Model no. 3-580: the inlet pressure regulator (IPR), Matheson Model no. 3075-1/4; the back-pressure regulator (BPR), Grove Valve Co's model no. 155 per Figure no. 11410-F-P2-A; and the Corblin A1C-250 metal-diaphragm compressor, serial no. 1430.

Valving consists of the following:

TYPE	CODE	SOURCE	PART NO.
Ball Valves	BV-1 to BV-10 BV-14 to BV-19	Whitey	SS-43S4
Ball Valves	BV-11 to BV-13	Whitey	SS-41S2
Check Valves	LV-1 to LV-3	Nu Pro	SS-2C-1
Check Valves	CV-0 to CV-4	Nu Pro	SS-4CA-3
Needle Valves	NV-1 and NV-4	Whitey	SS-31RS4
Needle Valves	NV-2 and NV-3	Whitey	SS-21RS4
3-Way Valves	3W-1 to 3W-3	Whitey	SS-43XS4
Sampling Valves	SV-1 to SV-3	Valco	4V6-HPaX with 100 μl coil ·
1/2 in. Gate Valves	GV-1 to GV-3	Any	

Key to Figure 1.

Figure 2 describes the spiral-wound cellulose acetate membranes used in this testing series. Other types of permeators, e.g., hollow-fiber type of either polyamide or polysulfone, were considered, but could not be obtained within the time constraints of this program. Membrane operation and the final geometry of the loaded housing are described in Figure 2. The effective area for the cellulose acetate membrane used during the study was ~ 1.7 ft^2. Membrane modules of other area sizes were either defective on receipt from the manufacturer or damaged during startup. This precluded obtaining scaling data for membrane area during the study.

Table II lists the feed gas compositions used for the study to determine the effects of feed composition on membrane performance. The different compositions are representative of diving gas mixtures used at various depths. Both the individual component permeabilities and the mixture permeabilities were evaluated at several different pressure drops up to 1200 psig feed pressure.

Table II
Feed Gas Composition

Component	Mix #1	Mix #2	Mix #3	Mix #4
	By Volume	Percent		
Helium	79.7	92.5	95.4	(58.0)
Nitrogen	15.9	5.9	3.6	(11.5)
Oxygen	4.3	1.6	1.0	(30.0)
Carbon Dioxide	0.075	0.027	0.017	(0.526)
Representative Depth (ft)	200.	600.	1000.	0 to 200
Expt'l Run Numbers	201-204	181-184	191-194	

Note: Experimental difficulties pre-empted the experiments using Mix #4.

Calculations: Permeability Constants, Separation Factors and Scaling of Results for System Design

The definition of a species' permeability in a given membrane is the product of its diffusivity and solubility in the membrane. It is this product which also results in the membrane's selectivity for one component of a mixture over another(9). However, since the permeability is actually a complicated function of the diffusivity and solubility, it is empirically measured and calculated by Equation 1.

$$q_i = k_i \, A \, \Delta p_i \tag{1}$$

where:

q_i = flow of component through the membrane
A = membrane area

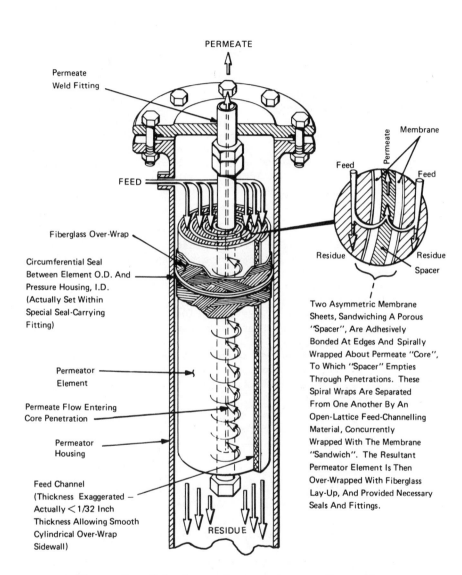

Figure 2. Spiral-wound permeator configuration.

p_i = partial pressure of component i
k_i = permeability constant for component i

Table III describes the permeability constants as determined by this set of experiments.

As Figure 3 shows, strict application of Equation 1 results in 11 equations in 11 unknowns and an iterative solution for the design sequence. To circumvent this cumbersome calculation, the system permselectivity was used to estimate the number of stages required in the design sequence for the separation, and Equation (1) was then used to estimate the area requirements knowing the stage outlet compositions of the permeate and residual streams.

In addition, separation factors were selected as the method of scaling the data, rather than the permeability coefficients, because this method provides a more reliable fit of the limited data available. The data was limited because the 58% helium mixture in Table II was not studied experimentally due to plasti-cization by water vapor of the last workable membrane module. In addition to being the most common helium mixture which the proto-type would encounter, this mixture represents lower values of the ratio β = [He]/[component i] than studied experimentally. Using the separation factors, the values of $\beta_{feed}/\beta_{permeate}$ for the 58% helium mixture lie between the origin (0,0), a fixed boundary, and the existing data when plotting $\beta_{feed}/\beta_{permeate}$. Use of permeabilities would require extrapolation as opposed to separa-tion factor interpolation. Therefore, the separation factor was selected for scaling.

The permselectivities, or separation factors, defined by Equation (3) were used to scale the experimental data for design purposes. The system permselectivity is analogous to the distil-lation separation factor

$$[\text{Sep. factor}]_{\chi}^{He} = \frac{(p_{He})_{feed}}{(p_x)_{feed}} \Big/ \frac{(p_{He})_{permeate}}{(p_x)_{permeate}} \tag{3}$$

discussed by McCabe & Smith(3) and by Treybal(5), defined by Equation (4), the relative volatility αij, for a binary system.

$$\alpha_{ij} = p_i^{\circ}/p_j^{\circ} = (p_A/p_B)_{vapor}/(p_a/p_B)_{liquid} \tag{4}$$

The permselectivity for membrane separations can also be calcu-lated by substituting fugacities calculated from an equation of state, here using the Beattie-Bridgeman equation, into Equation (3) for the partial pressure values (4). The values of the perm-selectivities in Table IV are relatively constant at a fixed feed composition in agreement with the approximately linear behavior noted in Figures 9-11.

Table III. Permeability Results

Run No.	① Dalton K_{He} @ Δp_{He}	K_{N_2} @ ΔN_2	$\dfrac{K_{He}(\Delta p_{He})}{K_{N_2}(\Delta p_{N_2})}\Big\} Q_1$	③ Permeate Concentrations % [He]	% [N₂]	$\dfrac{[He]}{[N_2]}\Big\} Q_2$	Comments
181	8.25 @ 164.8	0.44 @ 20.4	151.6	99.21	0.66	150.3	92.8% He, 5.63% N₂ feed gas
182	8.01 @ 337.5	0.59 @ 31.4	145.9	99.20	0.68	145.9	
183	7.72 @ 510.5	0.71 @ 42.5	130.6	99.09	0.76	130.4	
184	7.43 @ 682.9	1.25 @ 52.5	77.3	98.44	1.27	77.5	
185	6.31 @ 859.4	2.94 @ 59.0	31.3	96.13	3.08	31.2	
191	8.03 @ 169.5	(0.48) @ (13.4)	(211.6)	(99.50)	(0.47)	(211.7)	( ) Inaccurate because of failure to separate [O₂] from [N₂] in analysis
192	8.05 @ 353	0.72 @ 20.7	190.7	99.44	0.52	191.2	
193	7.7 @ 529.7	0.78 @ 28.0	186.8	99.37	0.53	187.5	95.2% He, 3.71% N₂ feed gas
194	7.2 @ 704.2	0.79 @ 35.3	181.8	99.37	0.55	180.7	
201	5.75 @ 266.4	0.46 @ 86.0	38.4	95.77	2.54	37.7	80.0% He, 15.6% N₂ feed gas
202	5.9 @ 417.1	0.56 @ 116.8	37.6	95.86	2.56	37.4	
203	6.2 @ 567.7	0.65 @ 147.7	36.7	95.83	2.61	36.7	
204	6.08 @ 715.5	0.85 @ 176.8	28.9	94.87	3.28	28.9	

Q_1 from Permeation Rates calculated by Dalton's Law
Q_2 from ratio of Permeate Concentrations

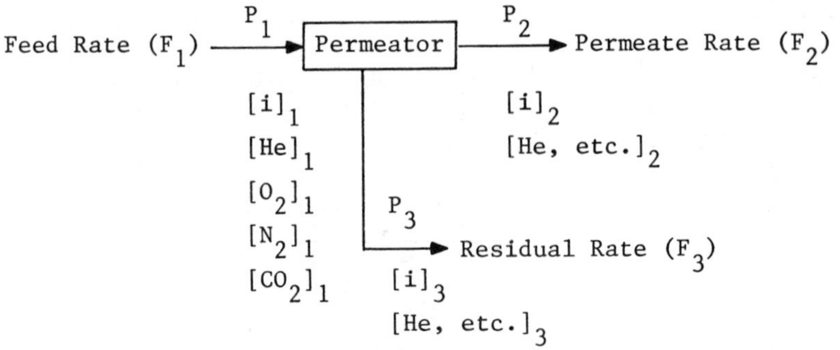

Feed Rate (F_1) $\xrightarrow{\quad P_1 \quad}$ [Permeator] $\xrightarrow{\quad P_2 \quad}$ Permeate Rate (F_2)

$[i]_1$ $[i]_2$

$[He]_1$ $[He, etc.]_2$

$[O_2]_1$

$[N_2]_1$ $\xrightarrow{\quad P_3 \quad}$ Residual Rate (F_3)

$[CO_2]_1$ $[i]_3$

            $[He, etc.]_3$

where P_1, P_2, P_3 = Pressure of F_1, F_2 and F_3 streams, respectively
and $[i]_1$, $[i]_2$, $[i]_3$ = concentration of i-<u>th</u> component in respective streams

Overall Material Balance, $F_1 = F_2 + F_3$
Component Material Balances, $F_1[i]_1 = F_2[i]_2 + F_3[i]_3$

Flux Equations, $F_2[i]_2 = k_i A \left(\dfrac{[i]_1 P_1 + [i]_3 P_3}{2} - [i]_2 P_2 \right)$

% Recovery $= \dfrac{[i]_2 F_2}{[i]_1 F_1} \times 100$

To perform design analysis for Helium Reclaimer:

Solve for	Using
4 Permeate Components	4 Flux Equations
4 Residual Components	4 Component Material Balances
1 Permeate Rate	1 Overall Balance
1 Residual Rate	1 Sum of Compositions
1 Membrane Area	1 Definition of Recovery

Total
11 Unknowns   11 Equations

Figure 3. Single-stage schematic and material balances.

Experimental Results and Discussion

To verify the membrane integrity prior to attempting separations, pure gas permeation rates for nitrogen and helium were determined and compared to the vendor's data supplied with the membrane. Figure 4 and Table V verify the vendor's data reasonably well for the only membrane which survived shipment and startup. The agreement of the nitrogen values is particularly indicative of the membrane's integrity.

Table IV
Separation Factors

Feed Composition			Values Calculated from Fugacity	
			Runs 181-184	
He	92.83	mole %	ΔP = 189 psi	α = 9.59
CO_2	.051		ΔP = 380 psi	α = 9.48
O_2	1.49		ΔP = 566 psi	α 8.34
N_2	5.63		ΔP = 751 psi	α 4.87
H_2O	.008			

Feed Composition		Runs 191-194	
He	95.23	ΔP = 187 psi	α = 9.86
CO_2	0.013	ΔP = 381 psi	α = 8.80
O_2	1.06	ΔP = 568 psi	α = 7.84
N_2	3.71	ΔP = 751 psi	α = 7.85
H_2O	.014		

Feed Composition		Runs 201-204	
He	79.96	ΔP = 376 psi	α = 5.64
CO_2	.076	ΔP = 568 psi	α = 5.78
O_2	4.34	ΔP = 756 psi	α = 5.75
N_2	15.60		
H_2O	0.22		

Permeate Composition Ratios	Runs 181-184 92.8% He	Runs 191-194 95.2% He	Runs 201-204 80% He
at ΔP = 380 psi			
He/sum	124	178	23
He/CO_2	5221	19888	1570
He/O_2	992	1421	60
He/N_2	144	191	38
ΔP = pressure range	189- 751	187- 751	376- 941
He/sum	126- 63	199- 158	23- 18
He/CO_2*	5221-3786	14214-11041	1570-1843
He/O_2	1044- 365	1421- 2481**	53- 60
He/N_2	150- 78	212- 181	28- 38

* first value of 3543 at 189 ΔP is anomalously low
** α increases as ΔP increases

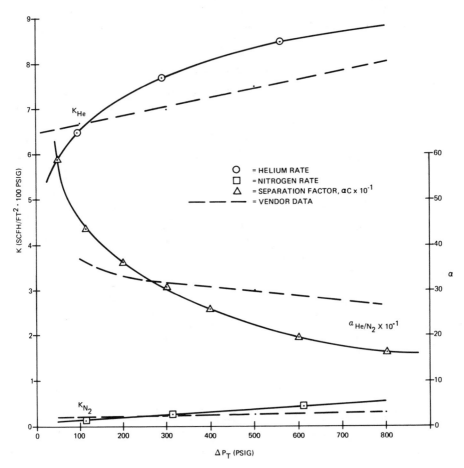

Figure 4.   Pure gas permeation rates, Permeator Element No. 0003.

Table V. Comparative Permeability Rate Data – WEC Vs Vendor

Element	Run	Gas	WEC Data			Vendor Data			Ratios		Comments
			ΔP	K_w	a_w	ΔP	K_v	a_v	K_w/K_v	a_w/a_v	
No. 0003	171B	He	95	6.6	44.0	100	6.7	35.3	1.0	1.2	All ratios close to unity. Signifies intact membrane. Later confirmed by good He/N$_2$ separation performance.
	161B	N$_2$	115	0.15		100	0.19*		0.8		
	173B	He	557	8.5	19.8	575	7.6*	29.2	1.1	0.7	
	163B	N$_2$	612	0.43		575	0.26*		1.6		

* = extrapolated value

Figure 5 shows that, during the experiments, helium recovery
varied linearly with pressure drop for all three test mixtures.
The 92% helium mixture, which was studied first, is displaced in
the figure from its expected position (shown by the dotted line)
except for the final data point. This final reading was taken at
a pressure drop much higher than the vendor tests. This irrever-
sibly compressed the membrane, altering its separation characteristics from the first four data points using 92% mixture,
as verified by subsequent experiments on the 80% and 95% helium
mixtures. Therefore, only the 80% and 95% mixture data were used
for the design studies.

Helium Enrichment: Permeabilities and Separation Factors

Enrichment of the preferred gas, helium in this case, is
most productively displayed as a plot of the feed concentration
of this gas relative to the remainder gases (or to a less
preferred gas in the mixture) versus the equivalent ratio in the
permeate. This is thus a plot of feed ratio vs. flux ratio. If
this is done for several feed compositions representing a significant preferred-gas composition range, enrichment curves can be
generated, such as those displayed in Figures 6 and 7. Here it
can be seen that the lower pressure differentials are more
desirable from a permeate quality standpoint.
It can further be seen that helium is least efficiently
enriched relative to carbon dioxide than to any other component.
It is, therefore, enrichment with respect to CO_2 that becomes
limiting and determines the number of stages necessary to achieve
the purity levels set by Table I. In this respect, selection of
a $\sim$ 380 psi ΔP condition appears particularly favorable, since it
yields the highest $[He]/[CO_2]$ enrichment.
Since these separation factors are the basis of the data
scale-up procedure, Figures 8, 9, and 10 display the separation
ratios for the 380 ΔP experimental data selected as the system
design point.
In these experiments, the measured helium flux through the
membrane was less than the flux predicted on the basis of the
average bulk concentrations. Consequently, the helium permeability coefficients calculated from observed membrane flux and
the bulk partial pressures are lower than the pure gas values
obtained by the membrane supplier or independently by us. At the
same time, observed nitrogen coefficients are higher than
predicted.
Concentration polarization may be a possible explanation for
this erratic permeability coefficient behavior; however, the
influence of concentration polarization in gaseous exchange
regimes such as the one herein reported is quite doubtful and
subject to dispute(7). Alternative possibilities include (a)
permeator functional loss due to on-going membrane compaction,

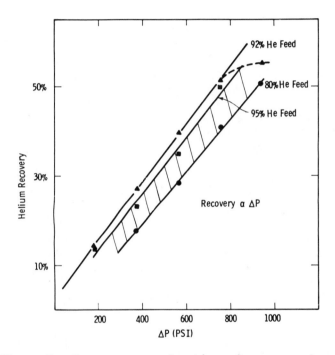

Figure 5. Recovery as a function of pressure drop.

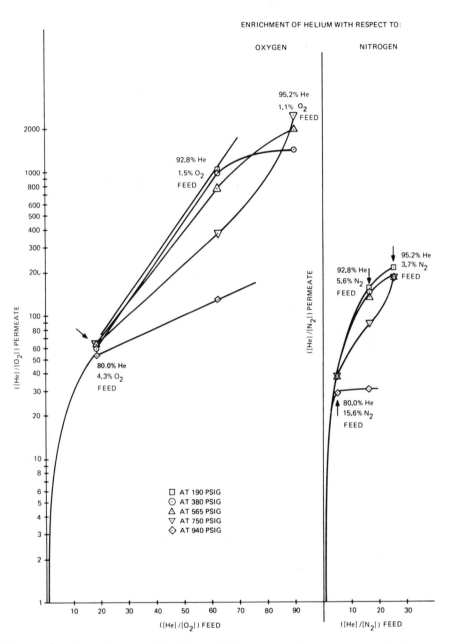

Figure 6.   Enrichment of helium with respect to oxygen and nitrogen.

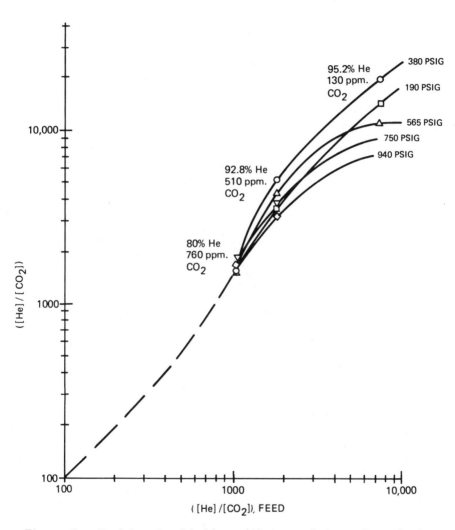

Figure 7. Enrichment of helium with respect to carbon dioxide.

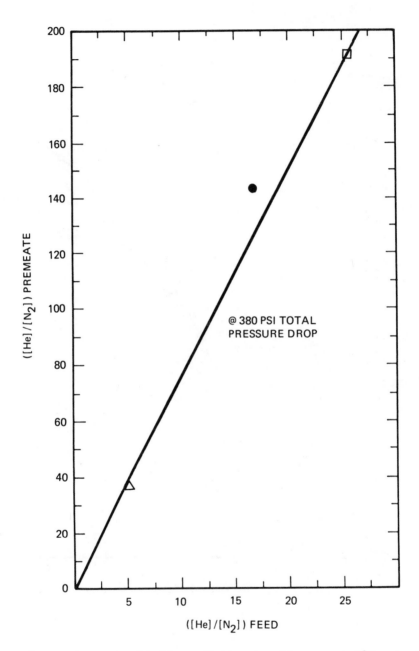

Figure 8.   Enrichment of helium relative to nitrogen at 380 psig ΔP.

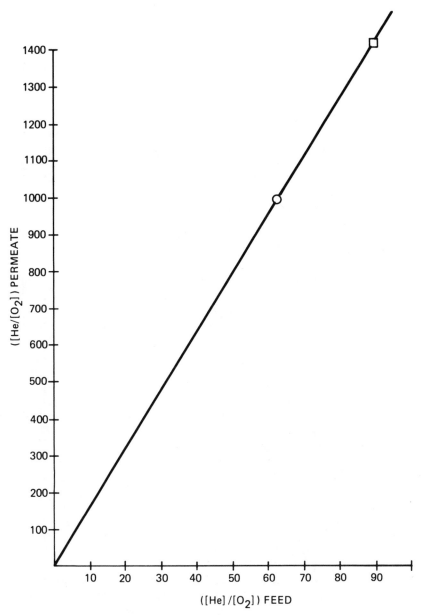

Figure 9.   Enrichment of helium relative to oxygen at 380 psig ΔP.

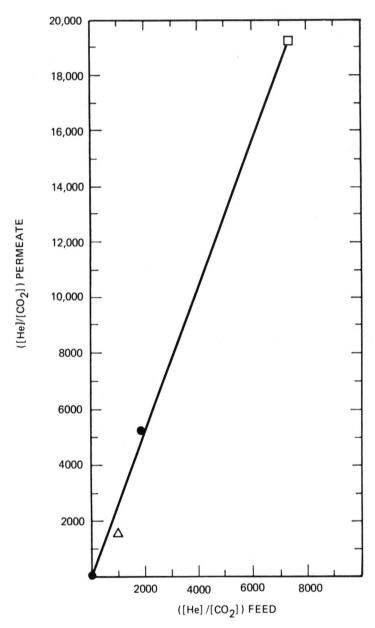

Figure 10. Enrichment of helium relative to carbon dioxide at 380 psig ΔP.

(b) loss due to some other factor such as gradual membrane
plasticization by water vapor, or (c) non-ideality of mixtures
especially at the higher pressures. The source of discrepancy
remains an issue open to future evaluations.

Hypothetical System Design

Table VI
Design Assumptions for 58% Helium Mixtures

1. Recovery α Area
    α ΔP (shown by data)
2. Linear Extrapolation of Permselectivities and Permeabilities
3. 50% Decrease in Flux due to Humidity Levels ≥ 80% (see
   Reference 8).

     In fulfillment of the contract requirements with the Navy, a
hypothetical helium recovery system was designed based on the
limited data available from the test program. These calculations
were performed to put the test results in perspective. They are
not, however, intended to describe a system that Westinghouse has
for market. Westinghouse has no plans to market such a system.
     Figure 11 and the accompanying material balances in Figure
12 show that five membrane separation stages are required in the
hypothetical system to meet the specified levels in Table I for a
58% helium mixture.
     Since the experimental data showed that the cellulose acetate
membranes concentrate water, the water must be removed from the
feed stream before it enters the permeation stages in twin
zeolite absorption columns. The dewatering system would be
backed up by high efficiency particulate filters rated to ∿ 0.3 μ.
Since the final product is to be breathable, diaphragm compressors would be used throughout the system to prevent gaseous
contamination by oil. A gas chromatograph would be used to
monitor effluent quality, while the pressure, humidity and temperature taps check the system operation as the probes in Figure
11 indicate.
     As Figure 12 shows, the fifth stage is required to meet the
CO_2 spec when treating the 58% helium mixture, while only four
stages are adequate to meet the target of ≥ 99.8% He purity.

Conclusions

     1. The design proposed here is a hypothetical preliminary
design assembled to determine the technical feasibility of using
semipermeable membranes to recover helium.
     2. More detailed data concerning the effects of humidity,
membrane area and lower helium concentrations in the feed stream
on helium recovery must be obtained before a final design can be
performed with confidence.

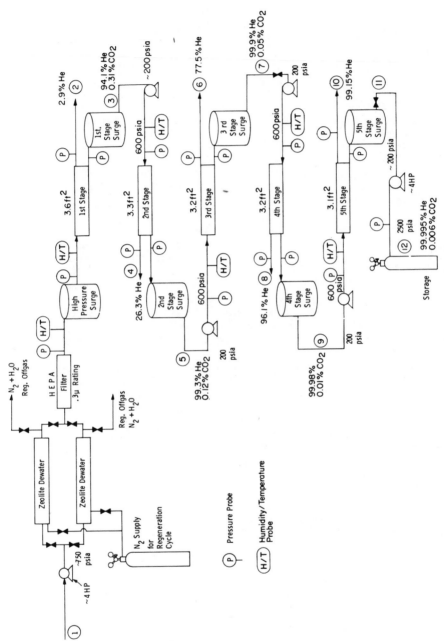

Figure 11. Helium reclaimer design.

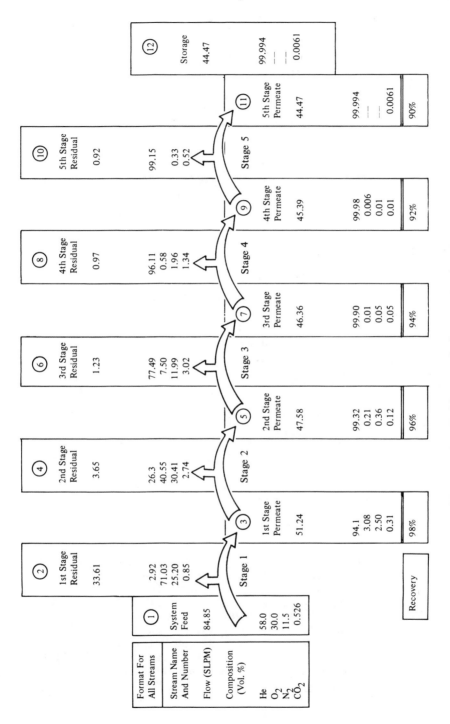

Figure 12. Helium reclaimer material balance.

3. For a hypothetical system design, based on the existing data, multistage systems are required to recover the helium from vented diving chamber gases in order to meet the design specifications outlined in Table I.
4. Five stages are required to achieve the helium purity goal and the carbon dioxide requirement in Table I.
5. Carbon dioxide is the limiting gas for the design as oxygen and nitrogen require only two stages to meet their design requirement levels.

Use of other membranes, having different properties, will, of course, result in a system different from that indicated above. Depending on the choice of membrane material, fewer or more, smaller or larger stages would be required for the system. The expected cost of reclaimed helium would be correspondingly affected.

Nomenclature

k	permeability coefficient (STD ft^3/ft^2 hr 100 psi)
p	pressure (atm)
T	temperature (°K except as noted)
ρ	density (g/cc)
L	some characteristic length (cm)
$[i]_2$	permeate concentration mole %
$[i]_3$	residual concentration mole %
$[i]_1$	feed concentration mole %
A	area
F_3	residual flow rate (SLPM)
F_1	feed flow rate (SLPM)
F_2	permeate flow rate (SLPM)
p_i^o	vapor pressure of component i
x_{ij}	ideal separation for distillation = p_i^o/p_j^o
α_{ij}	separation factor or permselectivity

Literature Cited

1. Howland, H. R.; Hulm, J. K. The Economics of Helium Conservation, Final Report to Argonne National Laboratory, Contract No. 31-109-38-2820, December,1974.
2. Helium Reclaimer Study, Westinghouse Electric Corporation, Proposal No. Y7420, April.1978.
3. McCabe, W. G.; Smith, I.C. Unit Operations of Chemical Engineering, McGraw-Hill, 1967, Second Edition.
4. Balzhizer, R. W.; Samuels, M. R.; Eliassen, J. D. Chemical Engineering Thermodynamics, Prentice Hall, 1972.
5. Treybal, R.E. Mass Transfer Operations, McGraw-Hill, 1968.

6. Hwang, S. J.; Kemmermeyer, K. Membranes in Separation, Wiley & Sons, 1975.

7. Bird, R. B.; Stewart, W. E.; Lightfoot, E.M. Transport Phenomena, Wiley & Sons, 1960.

8. Schell, W. J. "Membrane Applications to Coal Conversion Processes," Envirogenics Systems Company, October, 1976, NTIS No. FE2000-4.

9. Li, N. N., Ed.; Recent Developments in Separation Science, Vol. II, CRC Press, Cleveland, Ohio, 1972.

RECEIVED January 18, 1983

A Low-Temperature Energy-Efficient Acid Gas Removal Process

R. E. HISE and L. G. MASSEY—Consolidated Natural Gas Research
Company, Cleveland, OH 44106

R. J. ADLER, C. B. BROSILOW, and N. C. GARDNER—Case Western
Reserve University, Cleveland, OH 44106

L. AUYANG, W. R. BROWN, W. J. COOK, and Y. C. LIU—Helipump
Corporation, Cleveland, OH 44141

The CNG process removes sulfurous compounds, trace
contaminants, and carbon dioxide from medium to
high pressure gas streams containing substantial
amounts of carbon dioxide. Process features include 1) absorption of sulfurous compounds and
trace contaminants with pure liquid carbon dioxide, 2) regeneration of pure carbon dioxide with
simultaneous concentration of hydrogen sulfide and
trace contaminants by triple-point crystallization, and 3) absorption of carbon dioxide with a
slurry of organic liquid containing solid carbon
dioxide. These process features utilize unique
properties of carbon dioxide, and enable small
driving forces for heat and mass transfer, small
absorbent flows, and relatively small process
equipment.

Acid gas removal (AGR) refers to the separation of hydrogen
sulfide and carbon dioxide from gas mixtures. The simultaneous
removal of other trace sulfurous and nitrogenous compounds is
also usually implied, e.g., carbonyl sulfide, carbon disulfide,
mercaptans, hydrogen cyanide. A less emphasized but equally
important function of acid gas removal is the separation of
carbon dioxide from hydrogen sulfide and other trace contaminants. AGR thus serves to increase the market value of a gas by
removing diluents and undesirable compounds, to cleanse a gas of
compounds detrimental to downstream processing (catalyst poi-
sons), to separate pollutant free carbon dioxide for by-product
use or atmospheric venting, and to concentrate hydrogen sulfide
for sulfur recovery.

The major applications of AGR appear to be: (1) treating
natural gas containing sulfurous compounds and carbon dioxide,
(2) treating shift gases in substitute natural gas production
from coal, (3) treating synthesis gas made by partial oxidation

0097–6156/83/0223–0027$06.00/0

of sulfur-contaminated feedstocks, e.g., gasoline production
from coal via synthesis gas (Fischer-Tropsch reaction or through
the intermediates methanol or dimethyl ether), (4) acid gas
removal in hydrogen production from coal or residual feedstocks
for uses such as oil shale refining, and (5) treating synthesis
gas made from methane and steam (production of ammonia, hydrogen,
and oxo compounds). In short, acid gas removal is a key step in
the up-grading of natural gas and synthesis gas made from natural
gas, petroleum, or coal, and acid gas removal is an essential step
in most envisioned fossil fuel energy conversion processes.

A proprietary acid gas removal process is under development
by the CNG Research Company, a Subsidiary of the Consolidated
Natural Gas Company; most of the research has been done by
Helipump Corporation at Case Western Reserve University, Cleveland, Ohio. The goal is the development of a commercially
attractive process. The process separates carbon dioxide,
sulfurous molecules, and other trace contaminants from gas
mixtures containing low molecular weight components such as
hydrogen, carbon monoxide, methane, and nitrogen. The process
operates at temperatures down to about -80°C, and is especially
suited for medium to high pressure (300-1500 psia) gas mixtures
containing substantial amounts of carbon dioxide. This paper
provides background and motivation for alternative AGR technology, a description of the CNG process and its features, and
reviews the present status of process development.

Background and Motivation for Alternative AGR Technology

There are two major types of acid gas removal processes,
chemisorption and physical absorption. Typical chemisorption
agents are amines and hot potassium carbonate; typical physical
absorbents are dimethyl ether of tetraethylene glycol (Selexol)
and cold methanol (Rectisol). At low acid gas partial pressure,
less than about 200 psia, chemisorption is generally favored; at
acid gas partial pressures above 200 psia, physical absorption is
generally favored (1). The following discussion of AGR process
characteristics is oriented primarily toward treatment of crude
coal gasifier gases with pressure and composition favoring
physical absorption AGR.

Existing AGR processes envisioned for use in the production
of synthetic fuels from coal face unique challenges because these
AGR processes were developed primarily in response to the needs
of the petroleum and natural gas industries where crude gas
mixtures are relatively well-defined. In contrast to crude gas
mixtures in the petroleum and natural gas industries, crude coal
gasifier gas generally contains much more carbon dioxide, a much
higher ratio of carbon dioxide to hydrogen sulfide, and many
trace contaminants. The AGR step in synthetic fuels production
from coal must be capable of performing two tasks: (1) separation

of acid gases and the trace contaminants from the crude gas, and (2) separation of highly purified carbon dioxide from the hydrogen sulfide and trace contaminants.   Capital and energy cost estimates for AGR in synfuels production are high because processing and environmental requirements demand that these separations be sharp.   In fact, AGR is estimated to be the single most costly part of a substitute natural gas plant (excluding the oxygen plant), more costly than the gasification step or the methanation step ($\underline{2}$).

Existing physical absorption AGR processes are relatively energy inefficient for application in coal gasification; they use substantial amounts of steam or stripping gas to regenerate lean solvent and power to pump lean solvent into the AGR absorber. In the treatment of crude gas with substantial carbon dioxide content, work available by expansion of separated carbon dioxide from its partial pressure in the crude gas, typically 100-300 psia, to atmospheric pressure, is not recovered. In theory, an AGR process could recover and utilize this potential energy.

Existing AGR processes have difficulty treating crude gases with high ratios of carbon dioxide to hydrogen sulfide to produce a sulfurous stream in sufficient concentration for conversion to elemental sulfur in a Claus plant.   Although a Claus feed of 40% H_2S or greater is most desirable ($\underline{3}$), AGR processes treating crude coal gasifier gases typically produce an acid gas stream containing 25% H_2S or less ($\underline{4}$).   Separation of pure carbon dioxide for by-product use or for rejection to the atmosphere (environmental restrictions are typically 200-250 ppm total sulfur, 10 ppm as H_2S) is also difficult to achieve with existing AGR processes. Attempting to increase the hydrogen sulfide concentration of the acid gas stream and to attain high purity carbon dioxide with existing AGR processes usually results in steeply rising capital and operating costs.

The many trace contaminants in gases produced from coal include sulfurous compounds such as carbonyl sulfide, carbon disulfide, mercaptans, thiophene, and nitrogen compounds such as ammonia and hydrogen cyanide, and aliphatic and aromatic hydrocarbons having a broad range of volatility. Existing AGR processes have difficulty removing all of these trace contaminants to the low levels desired for downstream processing and carbon dioxide venting.   They also face potential technical and economic problems in solvent recovery and regeneration, e.g., solvent loss with treated or vent gases (methanol), and increased nitrogen or steam stripping for effective removal of contaminants.   The problem of trace contaminant removal and rejection is receiving increased attention in the evaluation and selection of AGR processes for proposed coal gasification plants sited in the United States ($\underline{5}$).

AGR Process Development Goals

The following goals were established at the start of efforts
to develop an acid gas removal process more suited to gas cleaning
in the production of synthetic fuels from coal, particularly
medium BTU and substitute natural gases:

- Low energy consumption
- Reduced capital cost
- Concentrated hydrogen sulfide for conversion to sulfur
- Pure carbon dioxide by-product
- Non-polluting (pure carbon dioxide vent gas)
- Removal and rejection of trace contaminants
- Non-corrosive with ordinary materials of construction.

Theoretical analysis indicated that the first goal, reduced
energy consumption, was realistic. A reversible acid gas removal
process separating acid gases from a 1000 psia crude gas
containing 30 mol % CO_2 would reject carbon dioxide at ambient
temperature and pressure, and be a net energy producer.

Process Description and Features

The CNG acid gas removal process is distinguished from
existing AGR processes by three features. The first feature is
the use of pure liquid carbon dioxide as absorbent for sulfurous
compounds; the second feature is the use of triple-point crystallization to separate pure carbon dioxide from sulfurous
compounds; the third feature is the use of a liquid-solid slurry
to absorb carbon dioxide below the triple point temperature of
carbon dioxide. Pure liquid carbon dioxide is a uniquely
effective absorbent for sulfurous compounds and trace contaminants; triple-point crystallization economically produces pure
carbon dioxide and concentrated hydrogen sulfide; for bulk carbon
dioxide absorption the slurry absorbent diminishes absorbent flow
and limits the carbon dioxide absorber temperature rise to an
acceptable low value. The sequence of gas treatment is shown in
Figure 1, an overview of the CNG acid gas removal process.

Carbon dioxide plays a central role in the CNG process both
as a pure component and in mixture with other compounds. The
triple point of carbon dioxide is referred to frequently in the
following discussion; it is the unique temperature and pressure
at which solid, liquid, and vapor phases of carbon dioxide can
exist at equilibrium (-56.6°C, 5.1 atm). The carbon dioxide
triple point is shown in Figure 2, a phase diagram for carbon
dioxide.

Precooling, Water Removal. The raw gas is cooled and
residual water vapor is removed to prevent subsequent icing. The

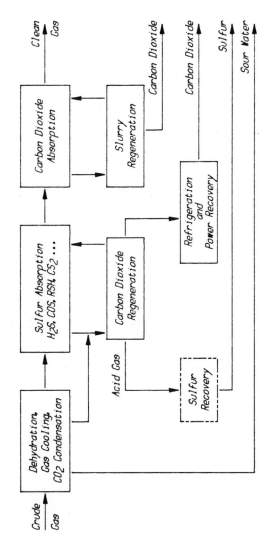

Figure 1.   CNG process overview.

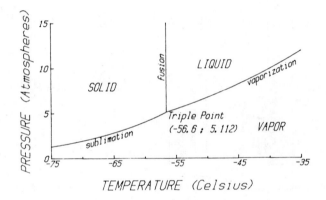

Figure 2.   Carbon dioxide phase diagram.

method of water removal depends on the crude gas composition. If
the crude gas is relatively free of C_2+ hydrocarbons, a regenerable molecular sieve can be used; otherwise, water can be removed
by solvent washing, e.g., dimethyl formamide, with dry solvent
regenerated by distillation. The economics of the two methods
appear competitive. The water-free crude gas is further cooled
to its carbon dioxide dew point by counter current heat exchange
with return clean gas and separated carbon dioxide.

The dew point of the gas depends on the gas pressure and the
carbon dioxide partial pressure. At fixed carbon dioxide
composition, the dew point is lowered as total pressure decreases; at fixed total pressure, the dew point is lowered as
carbon dioxide composition decreases. Calculated dew points for
a synthesis gas with 30 mol % carbon dioxide, for both ideal and
real gas behavior, are shown in Figure 3 for syn gas pressures up
to 1500 psia.

The dew point must be warmer than $-56.6^{\circ}C$ to permit use of
liquid carbon dioxide absorbent because pure liquid carbon
dioxide cannot exist below the triple point. The carbon dioxide
partial pressure, i.e., gas phase CO_2 mol fraction times total
pressure, of synthesis gas mixtures with $-56.6^{\circ}C$ dew points is
plotted versus synthesis gas pressure in Figure 4. Increasing
the H_2:CO ratio at fixed total pressure decreases the carbon
dioxide partial pressure required for a $-56.6^{\circ}C$ dew point.
Liquid carbon dioxide can be used to absorb sulfur molecules for
any combination of gas pressure and carbon dioxide partial
pressure which lies above the curves of Figure 4.

Carbon Dioxide Condensation, Sulfurous Compound Absorption.
Carbon dioxide is condensed by cooling the gas from its dew point
to about $-55^{\circ}C$; the condensate is contaminated with sulfurous
compounds. Substantial amounts of carbon dioxide can be condensed if the dew point is relatively warm. A synthesis gas at 1000
psia with 30 mol % carbon dioxide has a dew point of about $-30^{\circ}C$,
and approximately 65% of the carbon dioxide condenses to a liquid
in cooling to $-55^{\circ}C$. Removal of carbon dioxide by condensation
reduces the amount which must be removed subsequently by absorption. Condensation is preferred over absorption because it is
more reversible and hence is more energy efficient and less
capital intensive.

The gas at $-55^{\circ}C$ is scrubbed clean of sulfurous compounds and
remaining trace contaminants using pure liquid carbon dioxide.
Liquid carbon dioxide has several desirable properties as an
absorbing medium for sulfurous compounds. Surprisingly, liquid
carbon dioxide absorbs carbonyl sulfide (COS) more effectively
than it absorbs hydrogen sulfide as shown by their relative
vaporization coefficients, i.e., $K_{COS} < K_{H_2S}$ ($K_i \equiv y_i/x_i$, $y_i =$
vapor mole fraction component i, $x_i =$ liquid mole fraction
component i). Thus, an absorber for sulfurous compounds designed
to remove hydrogen sulfide using liquid carbon dioxide will also

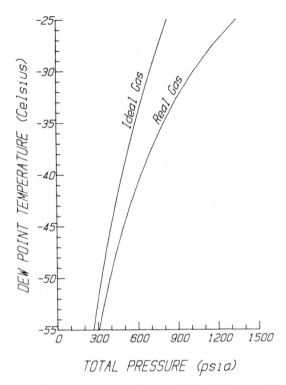

Figure 3.   Dew point vs. pressure, synthesis gas with 30% CO_2.

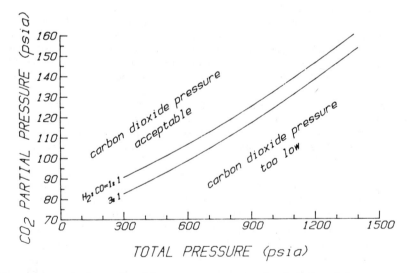

Figure 4.   Carbon dioxide pressures required for sulfur compound absorption.

remove COS and all other less volatile sulfurous compounds in the gas. The opposite is true concerning COS for both cold methanol (Rectisol) and the dimethyl ether of polyethylene glycol (Selexol solvent). For comparison, the ratio $K_{H_2S}:K_{COS}$ is plotted in Figure 5 versus liquid phase composition from 100 mol% methanol to 100 mol% carbon dioxide. The ratio $K_{H_2S}:K_{COS}$ for Selexol solvent is even lower than for methanol.

The physical properties of liquid carbon dioxide also enhance its use as an absorbent in general. The low viscosity (about 0.35 centipoise at -55^OC) and high density of liquid carbon dioxide promote high stage efficiency. For comparison, the viscosity of methanol is 7 to 8 times greater over the temperature range 0 to -55^OC (Figure 6), and Selexol solvent's viscosity is 5 to 10 centipoise around room temperature. Liquid carbon dioxide is denser (specific gravity 1.17 at -55^OC) than most absorbents which, together with its low viscosity, enables high liquid and gas flow rates in absorption and stripping towers. For comparison, methanol has a specific gravity of about 0.85 (-40^OC), and Selexol solvent about 1.03 at room temperature. Carbon dioxide has a relatively low molecular weight of 44 compared to Selexol solvent's molecular weight of over 200 (6). Low molecular weight favors high gas solubility per unit volume of solvent. A comparison of physical properties for methanol, Selexol solvent, and liquid carbon dioxide is summarized in Table I. The liquid carbon dioxide used to absorb the sulfurous compounds and other trace impurities is not purchased but rather is obtained from the crude gas being treated.

The liquid carbon dioxide absorbent, with sulfurous compounds, other trace contaminants, and perhaps some co-absorbed light hydrocarbons such as methane and ethane, is combined with the contaminated liquid carbon dioxide condensed in precooling to -55^OC. The combined carbon dioxide stream, typically 3 to 5 mol % hydrogen sulfide, is stripped of light hydrocarbons if necessary, and sent to the carbon dioxide regenerator. The treated gas, containing less than 1 ppm H_2S, leaves the sulfur absorber at essentially -55^OC with carbon dioxide the only significant impurity yet to be removed.

Carbon Dioxide Regeneration by Triple Point Crystallization.
Crystallization of carbon dioxide produces pure carbon dioxide (less than 1 ppm sulfur) and concentrated hydrogen sulfide (up to 75 mol %). The great advantage of crystallization is that pure solid carbon dioxide crystals form from mother liquor containing sulfurous and various other trace compounds. In contrast, separation of pure carbon dioxide from hydrogen sulfide by distillation is difficult because of the very small relative volatility between carbon dioxide and hydrogen sulfide at low hydrogen sulfide concentrations (7). Solid carbon dioxide is not known to form solid solutions with any of the trace compounds likely to be absorbed by liquid carbon dioxide in the sulfurous

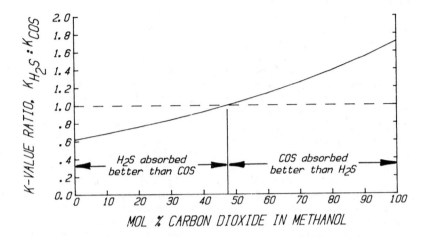

Figure 5.   Equilibrium K ratios.

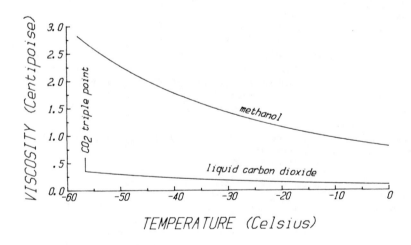

Figure 6.   Viscosity, methanol, and liquid carbon dioxide.

Table I

Properties of Physical Absorbents

Solvent	MW	Sp.Gr.$^{T°C}$	$\mu(°C)$ centipoise	BP,°C	FP,°C
methanol	32	0.85^{-40}	1.8(-40)	64.8	-97.5
dimethyl ether of polyethylene glycol	200	1.03	6.0(+25)	-	-20 to -30
liquid carbon dioxide	44	1.17^{-55}	0.35(-55)	$-78.5^{\dagger}$	-56.6^{*}

†solid sublimation temperature
*triple point, 75.1 psia

compound absorber. Consequently, the CNG process sharply rejects trace contaminants with the hydrogen sulfide rich acid gas stream.

The crystallization process employed is direct-contact tri-ple-point crystallization with vapor compression. The crystal-lization process is a continuous separation cascade analogous to continuous distillation. The cascade operates at temperatures and pressures near the triple-point of carbon dioxide such that vapor, liquid, and solid phases coexist nearly in equilibrium. Solid carbon dioxide is formed by flashing; solid carbon dioxide is melted by direct contact with condensing carbon dioxide vapor. No heat exchange surfaces are necessary to transfer the latent heat involved. Crystal formation occurs at pressures slightly below the triple-point; crystal melting occurs at pressures slightly above the triple-point (several psi in each case). Carbon dioxide vapor is compressed from the crystal formation pressure to the crystal melting pressure. Only a few stages of recrystallization are required to achieve extremely pure carbon dioxide.

Pure carbon dioxide produced by triple-point crystallization is split into two streams, one is absorbent recycled to the sulfurous compound absorber, the other is sent back through the process for refrigeration and power recovery, and is delivered as a product stream or is vented to the atmosphere. The amount of excess liquid carbon dioxide produced at triple-point conditions can be substantial. For example, in treating a 1000 psia synthesis gas with 30 mol % carbon dioxide, nearly 40% of the total carbon dioxide to be removed from the syn gas is separated as pure triple-point liquid available for refrigeration and power recovery. The other product of triple-point crystallization, a concentrated hydrogen sulfide stream, also is recycled back through the process for refrigeration recovery before being sent to the sulfur recovery unit.

The triple-point crystallization of carbon dioxide is illus-trated in Figure 7, which shows a schematic carbon dioxide phase diagram expanded about the triple-point and a closed-cycle triple-point crystallizer operating with pure carbon dioxide. The operation of this closed-cycle unit is identical to that of a unit in the stripping section of a continous crystallizer cascade, except that in the cascade vapor would pass to the unit above, and liquid would pass to the unit below.

The adiabatic flash pressure P_f, maintained slightly below the triple-point pressure, causes liquid to spontaneously vapor-ize and solidify. The ratio of solid to vapor is determined by the heats of fusion and vaporization; for carbon dioxide about 1.7 moles of solid are formed for each mole vaporized. The solid, more dense than the liquid, falls through a liquid head and forms a loosely packed crystal bed at the bottom. The liquid head is about 10-12 feet, and increases the hydrostatic pressure on the solid to melter pressure P_m. The crystal bed depth is about two

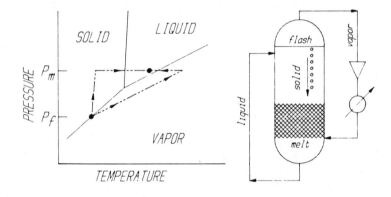

Figure 7.   Carbon dioxide, triple point crystallization.

feet. In actual operation with sulfur compounds present, a small
backwash flow is maintained upward through the bed to wash the
crystals and prevent mother liquor from penetrating into the bed.
The vapor is withdrawn from the flash zone, compressed to melter
pressure P_m, sensibly cooled to near saturation, and dispersed
under the solid bed where it condenses and causes solid to melt.
The liquid is withdrawn and returned to the flash zone with a pump
(not shown). The process cycle is traced on the phase diagram of
Figure 7.

Concentration changes observed between mother liquor in the
flash zone and liquid product in the melt zone of an experimental
triple-point crystallizer have been dramatic. A qualitative
concentration profile typical of those observed in the experimental unit is shown in Figure 8. The mother liquor concentration
is relatively uniform above the packed bed, but a sharp drop in
contaminant concentration occurs within the top several inches of
the loosely packed crystal bed. Concentration changes of the
order 500 to 5000 have been observed for representative sulfurous
compounds and trace contaminants, including hydrogen sulfide,
carbonyl sulfide, methyl mercaptan, ethane, and ethylene. Concentration profiles calculated for the packed bed of solid carbon
dioxide using a conventional packed bed axial dispersion model
agree very well with the observed experimental profiles.

Final Carbon Dioxide Removal. Carbon dioxide remaining in
the gas after sulfurous compound absorption at -55°C is absorbed
at temperatures below the carbon dioxide triple-point with a
slurry absorbent. The slurry absorbent is a saturated solution
of organic solvent and carbon dioxide containing suspended
particles of solid carbon dioxide. The low carbon dioxide
partial pressure of regenerated slurry absorbent, about one
atmosphere, is considerably less than the carbon dioxide partial
pressure of the entering raw gas, and provides the driving force
for carbon dioxide absorption. As carbon dioxide is absorbed,
the released latent heat warms the slurry and melts solid carbon
dioxide. The direct refrigeration provided by the melting of
solid carbon dioxide moderates the appreciable solvent temperature rise normally associated with bulk carbon dioxide absorption.

The large effective heat capacity of the liquid-solid slurry
absorbent enables relatively small slurry flows to absorb the
carbon dioxide heat of condensation with only modest absorber
temperature rise. This contrasts with other acid gas removal
processes in which solvent flows to the carbon dioxide absorber
are considerably larger than flows determined by vapor-liquid
equilibrium constraints. Large flows are required to provide
sensible heat capacity for the large absorber heat effects.
Small slurry absorbent flows permit smaller tower diameters
because allowable vapor velocities generally increase with reduced liquid loading (8).

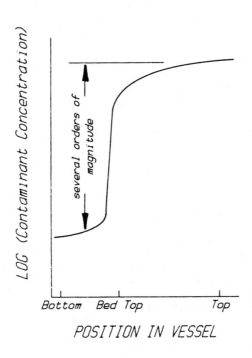

Figure 8.   Concentration profile.

The liquid solvent of the slurry absorbent is the sink for absorbed carbon dioxide vapor and melted solid carbon dioxide. The carbon dioxide rich solvent exits the absorber near the triple-point temperature but containing no solid carbon dioxide, and is stripped of methane and lighter molecules if necessary. The carbon dioxide rich absorbent is next cooled by external refrigeration and flashed in stages at successively lower pressures to generate a cold slurry of liquid solvent and solid carbon dioxide. The carbon dioxide flash gas is recycled back through the process for refrigeration and power recovery, and is delivered as a product stream or is vented to the atmosphere. The regenerated slurry absorbent is recirculated to the carbon dioxide absorber.

The absorption of carbon dioxide with slurry and the regeneration of slurry by flashing carbon dioxide require small temperature and pressure driving forces. The small driving forces derive from the huge surface area of the solid carbon dioxide particles and the low viscosity of the slurry. Compared with other sub-ambient temperature carbon dioxide removal processes, the CNG process requires less refrigeration even though process temperatures are often lower.

The solubility of solid carbon dioxide in several solvents (9,10) is shown in Figure 9. A preferred range of carbon dioxide solubility for slurry absorption is defined qualitatively by the two dashed lines of Figure 9. The indicated preferred solubility range derives from computer simulations of slurry absorption based on experimental solubility data for various solvents, and of hypothetical slurry absorption based on constant carbon dioxide partial pressure driving force per stage. Solvents with preferred solubility include ethers and ketones (11), e.g., di-n-butyl ether and methyl ethyl ketone. These solvents have a relatively linear change in solubility with temperature (as compared with hexane or methanol), low viscosity when saturated with carbon dioxide, low melting points, and low vapor pressures. The thermodynamic ideal solubility is shown for reference.

Carbon dioxide removal by slurry absorption is attractive down to about $-75^{\circ}C$, a temperature easily achieved by slurry regeneration to slightly above one atmosphere carbon dioxide pressure. For example, with a $-75^{\circ}C$ exit gas temperature, slurry absorption reduces the carbon dioxide content of a 1000 psia synthesis gas from about 13 to about 4 mole percent, a 70% reduction in carbon dioxide content. The exact level to which carbon dioxide can be removed from a treated gas by slurry absorption also depends on the solubility of solid carbon dioxide in the treated gas; the solubility of solid carbon dioxide in synthesis gas $(3H_2:CO)$ is illustrated in Figure 10 for several synthesis gas pressures. Fine removal of carbon dioxide to lower levels is accomplished by conventional absorption into a slip stream of the slurry solvent which is regenerated to meet particular product gas carbon dioxide specifications.

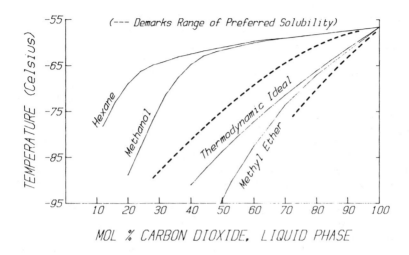

Figure 9. Solubility of solid carbon dioxide in selected solvents.

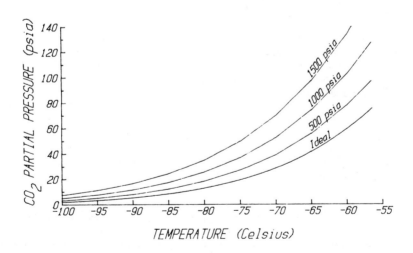

Figure 10. Solubility of solid carbon dioxide in synthesis gas.

Present Status

An extensive process data base has been accumulated which
includes: (1) vapor-liquid-solid equilibrium data for binary
systems of carbon dioxide with sulfurous compounds and with
ethers and ketones, and multicomponent systems including hydrogen, carbon monoxide, methane, carbon dioxide, and hydrogen
sulfide; (2) a bench-scale slurry formation and melting apparatus
which provides data on slurry pumping, and on rates of slurry
formation and melting, under various conditions of pressure
driving force and agitation; and (3) a bench-scale closed-cycle
triple-point crystallizer which provides data on rates of crystallization and melting, on the relation between these rates and
the pressure driving forces applied, and on separation factors
attainable in a single stage of crystallization.
The process features of carbon dioxide triple-point crystallization and slurry absorption of carbon dioxide have been
demonstrated with the first generation bench-scale apparatus.
Current efforts are focused on the design and construction of an
improved version of the carbon dioxide triple-point crystallizer
in cooperation with the U S Department of Energy. Future efforts
are planned to design and construct absorption units to study
multi-stage slurry absorption of carbon dioxide, and the more
conventional gas-liquid absorption of sulfuruous compounds with
liquid carbon dioxide.

Summary

The CNG process meets three objectives commonly acknowledged
by experts in gas purification to be highly desirable of acid gas
removal in synthetic fuels production from coal:

- production of concentrated hydrogen sulfide for conversion to sulfur in a Claus plant
- production of pure carbon dioxide for atmospheric venting, for enhanced oil recovery, or other uses
- removal of trace contaminants from the treated gas, and
  easy solvent regeneration with respect to trace contaminants

Although the process requires the treated gas to have a certain
minimum carbon dioxide partial pressure for removal of sulfurous
compounds with liquid carbon dioxide, promising new SNG processes
under development produce medium to high carbon dioxide content
crude gases ideally suited for acid gas removal via the CNG
process (12,13). The novel features of the CNG process have been
demonstrated with bench-scale process development units; second
generation process development units are in various stages of

planning, design, construction, and operation; readiness for commercial scale demonstration is anticipated in the time frame 1985-1987.

Literature Cited

1. Christensen, K.G., and W.J. Stupin, "Merits of Acid Gas Removal Processes," Hydrocarbon Processing, Feb. 1978, p. 125.
2. Factored Estimates for Western Coal Commercial Concepts, C.F. Braun and Company, Project 4568-NW, ERDA-AGA (1976).
3. Beavon, D.K., Kouzel, B., and Ward, J.W., "Claus Processing of Novel Acid Gas Streams," Symposium on Sulfur Recovery and Utilization, Division of Petroleum Chemistry, Inc., American Chemical Society, Atlanta Meeting, March 29-April 3, 1981.
4. Ghassemi, M., Strehler, D., Crawford, K., and Quinlivan,S., "Applicability of Petroleum Refinery Control Technologies to Coal Conversion," EPA Report No. EPA-600/7-78-190.
5. Rosseau, R.W., Kelly, R.M., and Ferrell, J.K., "Evaluation of Methanol as a Solvent for Acid Gas Removal in Coal Gasification Processes," Symposium on Gas Purification, AIChE Spring National Meeting, Houston, Texas, April 1981.
6. Sweny, J.W., "The Selexol Solvent Process in Fuel Gas Treating," paper presented at 81st National AIChE Meeting, Kansas City, Missouri, April 11-14, 1976.
7. Sobocinski, D.P., and Kurata, F., "Heterogeneous Phase Equilibria of the Hydrogen Sulfide - Carbon Dioxide System," AIChE J., Vol. 5, No. 4, p. 545, 1959.
8. King, C.J., Separation Processes, McGraw-Hill, p. 547, 1971.
9. Kurata, F., "Solubility of Solid Carbon Dioxide in Pure Light Hydrocarbons and Mixtures of Pure Light Hydrocarbons," GPA Research Report RR-10, Feb. 1974.
10. Baume, G., and F.L. Perrot, Recherches Quantitatives Sur Les Systemes Volatiles, Journal de Chimie Physique, 12, (1914).
11. Adler, R.J., Brosilow, C.B., Brown, W.R., and Gardner, N.C., "Gas Separation Process," U.S. Patent 4,270,937, June 2, 1981.
12. "Direct methanation promises lower gas costs," C & EN, p. 30, April 5, 1982.
13. Kaplan, L.J., ed., "Methane from coal aided by use of potassium catalyst," Chemical Engineering, p. 64, March 22, 1982.

RECEIVED December 23, 1982

Implications of the Dual-Mode Sorption and Transport Models for Mixed Gas Permeation

R. T. CHERN, W. J. KOROS, E. S. SANDERS, S. H. CHEN, and
H. B. HOPFENBERG

North Carolina State University, Department of Chemical Engineering,
Raleigh, NC 27650

The concept of unrelaxed volume in glassy
polymers is used to interpret sorption and
transport data for pure and mixed penetrants. A
review of recent sorption and permeation data for
mixed penetrants indicates that competition for
sorption sites associated with unrelaxed gaps
between chain segments is a general feature of
gas/glassy polymer systems. This observation
provides convincing support for the use of the
Langmuir isotherm to describe deviations from
simple Henry's law sorption behavior.
The observed flux depressions of a component
in a mixture, relative to its pure component value
at an equivalent partial pressure driving force,
derives from the above sorption competition
mechanism which influences the effective
concentration gradient driving diffusion across
the membrane. The competition among penetrants
for excess volume described above should be an
important consideration for modeling essentially
all permselective gas separation membranes.
Significant plasticizing effects may mitigate flux
reductions caused by the above competitive effects
at high pressures in plasticization-prone
polymers, but would also lead to selectivity
losses which are highly undesirable. The
permeation behavior of stiff-chain,
plasticization-resistant polymers which are likely
to comprise the next generation of gas separation
polymers will be appropriately treated by the
model discussed here.

0097–6156/83/0223–0047$07.75/0

   High membrane permselectivity is generally associated with
a rigid chain backbone of the constituent polymer, providing
sieving on a molecular scale (1,2). At high upstream driving
pressure (i.e., at high sorbed penetrant concentration),
plasticization of the polymer may reduce the permselectivity of
the membrane. Although plasticization will occur at
sufficiently high sorbed concentrations for essentially all
glassy materials, selection of a membrane material with a very
high glass transition temperature and extraordinary inherent
backbone rigidity should minimize effects associated with
plasticization.
   Often the ratio of pure component permeabilities at an
arbitrary pressure for the components of interest is used as an
indication of the potential selectivity of a candidate
membrane. While this approach offers a useful approximation,
the actual selectivity and productivity (flux) observed in the
mixed gas case may be surprisingly different than predicted on
the basis of the pure component data. As shown in Figure 1,
prior to the onset of plasticizing, the presence of a second
component B can depress the observed permeability of a
component A relative to its pure component value at a given
upstream driving pressure of component A. In the words of Pye
et al. (3), the permeability of a membrane to a component A may
be reduced due to the sorption of a second component B in the
polymer which ". . . effectively reduces the microvoid content
of the film and the available diffusion paths for the non
reactive gases". The present paper will offer additional
experimental and theoretical insight into this interesting
effect that is characteristic of glassy polymers.

Unrelaxed Volume – Relation To Sorption And Transport Properties Of Glassy Polymers

   The concept of "unrelaxed volume", V_g-V_ℓ, illustrated in
Figure 2 may be used to interpret sorption and transport data
in glassy polymers exposed to pure and mixed penetrants. The
extraordinarily long relaxation times for segmental motion in
the glassy state lead to trapping of nonequilibrium chain
conformations in quenched glasses, thereby permitting miniscule
gaps to exist between chain segments. These gaps can be
redistributed by penetrant in so-called "conditioning"
treatments during the initial exposure of the polymer to high
pressures of the penetrant (4). Following such initial
exposure to penetrant, settled isotherms characterized by
concavity to the pressure axis at low pressures and a tendency
to approach linear high pressure limits are observed as shown
in Figure 3. Reference to penetrant-induced conditioning
effects have been made continuously since the very earliest
investigations of penetrant/glassy polymer sorption

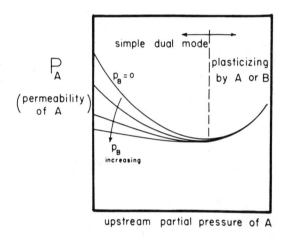

Figure 1. Permeability of a glassy polymer to penetrant A in the presence of varying partial pressures of penetrant B.

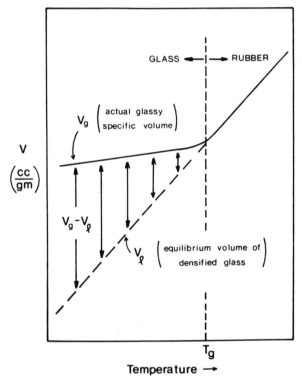

Figure 2. Schematic representation of the unrelaxed volume, $V_g - V_\ell$, in a glassy polymer. Note that the unrelaxed volume disappears at the glass transition temperature, T_g.

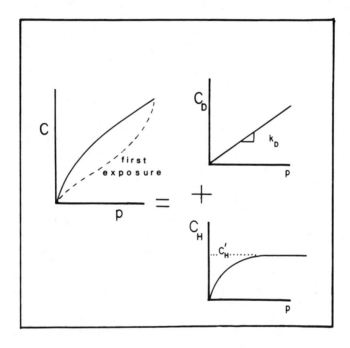

Figure 3.   Schematic representation of the dual mode sorption concept.

behavior ($\underline{2},\underline{4},\underline{5},\underline{6}$). If one "overswells" the polymer with a
highly sorbing penetrant such as a high activity vapor and then
removes the penetrant, the excess volume which is introduced
will tend to relax quickly at first, followed by a very slow,
long term approach toward equilibrium ($\underline{6}$).

At extremely high gas conditioning pressures, substantial
swelling of the polymer sample can occur with a resultant
increase in the value of $V_g - V_\ell$ ($\underline{5}$). For the case of
conditioning with gases such as CO_2 at pressures less than 30
atm, however, consolidation in the absence of penetrant
following the conditioning treatment is typically unmeasurable
since penetrant sorption uptake is not very extensive (< 4-5%
by weight) ($\underline{4},\underline{5}$). As a result, in such cases <u>redistribution</u> of
the originally present intersegmental gaps may be the primary
process occurring during the first exposure of the polymer to
high pressures of penetrant as shown in Figure 3 ($\underline{4}$).

An interpretation of the observed conditioning behavior
that occurs during the primary penetrant exposure in the
<u>absence</u> of large swelling effects may be couched in terms of
coalesence of packets of the original intersegmental gaps.
Redistribution of chain conformations, consistent with optimal
accomodation of the penetrant in the unrelaxed volume between
chain segments may permit this process during the conditioning
treatment. This rearrangement would tend to produce a more or
less densified matrix with a small volume fraction of
essentially uniformly distributed molecular scale gaps or
"holes" throughout the matrix. In such a situation, one can
appreciate the meaning of two slightly different molecular
environments in the glass in which sorption of gas may occur.
Consider first the limiting case in which a highly annealed,
truly equilibrium densified glass characterized by "V_ℓ" in
Figure 2 is exposed to a given pressure of a penetrant. In
this case, all gaps are missing, but there will clearly still
be a certain characteristic sorption concentration (C_D) typical
of true molecular dissolution in the densified glass, similar
to that observed in low molecular weight liquids or rubbers
(above Tg). Next, consider a corresponding conditioned
<u>nonequilibrium</u> glass (illustrated by "V_g" in Figure 2)
containing unrelaxed volume in which the surrounding matrix has
been more or less densified by the coalesence of gaps to form
molecular scale berths for penetrant. A local equilibrium
requirement leads to an average local concentration of
penetrant held in the uniformly distributed molecular scale
gaps (C_H) in equilibrium with the "dissolved" concentration
(C_D) at any given external penetrant pressure or activity.
This simple physical model can be described analytically up to
reasonably high pressues (generally for pressures less than or
equal to the maximum conditioning pressure employed ($\underline{4}$)) in
terms of the sum of Henry's law for C_D and a Langmuir isotherm
for C_H.

$$C = C_D + C_H \tag{1}$$

$$C = k_D p + \frac{C_H' b p}{1 + b p} \tag{2}$$

where k_D is the Henry's law constant that characterizes sorption of penetrant in the densified regions which comprise most of the matrix. The parameter b characterizes the affinity of the penetrant for the interchain gap regions in the polymer. The parameter C_H' is the Langmuir sorption capacity of the glassy matrix and can be interpreted directly in terms of Figure 2 in which the unrelaxed volume, $V_g - V_\ell$, corresponds to the summation of all of the molecular scale gaps in the glass. As shown in Figure 4, the Langmuir capacity of glassy polymers tends to approach zero in the same way that $V_g - V_\ell$, approaches zero at the glass transition temperature (see Figure 2). This qualitative observation can be extended to a quantitative statement in cases for which the effective molecular volume of the penetrant in the sorbed state can be estimated. As a first approximation, one may assume that the effective molecular volume of a sorbed CO_2 molecule is 80 Å^3 in the range of temperature from 25°C to 85°C. This molecular volume corresponds to an effective molar volume of 49 cc/mole of CO_2 molecules and is similar to the partial molar volume of CO_2 in various solvents, in several zeolite environments, and even as a pure subcritical liquid (See Table 1) (4,8). The implication here is not that more than one CO_2 molecule exists in each molecular scale gap, but rather that the effective volume occupied by a CO_2 molecule is roughly the same in the polymer sorbed state, in a saturated zeolite sorbed state and even in a dissolved or liquid-like state since all of these volume estimates tend to be similar for materials that are not too much above their critical temperatures. With the above approximation, the predictive expression given below for C_H' can be compared to independently measured values for this parameter from sorption measurements.

$$C_H' = \left[\frac{V_g - V_\ell}{V_g} \right] \rho^* \tag{3}$$

where ρ^* is the equivalent density of CO_2 (~1/49 mole/cc) discussed above. The comparison of measured (4,9-11) and predicted C_H''s calculated from Eq (3) using reported dilatometric parameters (12-15) for the various glassy polymers is shown in Figure 5. The correlation is clearly impressive.

Application of Eq (3) to highly supercritical gases is somewhat ambiguous since the effective molecular volume of sorbed gases under these conditions is not easily estimated. A

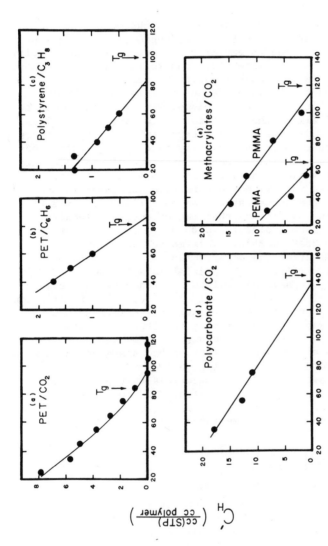

Figure 4. Langmuir sorption capacity, C_H', as a function of temperature for several polymer/penetrant systems. Note that C_H' disappears near T_g. (Reproduced with permission from Ref. 7. Copyright John Wiley & Sons, 1981.)

Table 1: Effective Molar or Partial Molar Volume of CO_2 in Various Environments at 25°C.

Environment	Molar or Partial Molar Volume (cc/mole)	Ref
Carbon Tetrachloride	48.2	8
Chlorobenzene	44.6	8
Benzene	47.9	8
Acetone	44.7	8
Methyl Acetate	44.5	8
4A or 5A Zeolite at saturation of capacity	52.4	4

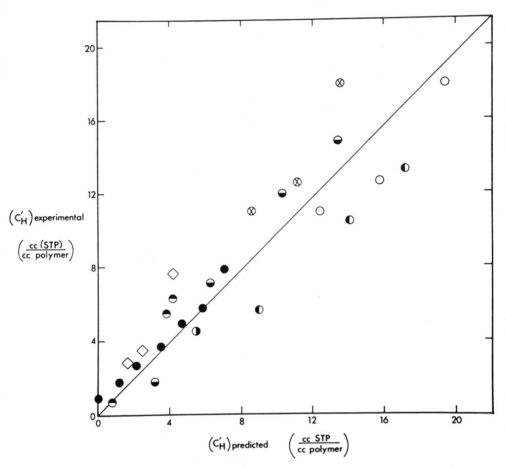

Figure 5. Quantitative comparison of experimentally measured
values of C_H' for CO_2 in various polymers with the predictions
of Equation 3. [●, poly(ethylene terephthalate) at 25, 35, 45,
55, 65, 75, and 85 °C], [◐, poly(benzyl methacrylate) at 30 °C],
[◑, poly(phenyl methacrylate) at 35, 50, and 75 °C], [◇, poly-
(acrylonitrile) at 35, 55, and 65 °C], [○, ⊗ polycarbonate at
35, 55, and 75 °C: the ○ symbols refer to predictions using
dilatometric coefficients from Ref. 14; the ⊗ symbols refer to
predictions using coefficients from Ref. 15]. [◒, poly(ethyl
methacrylate) at 30, 40, and 55 °C], [◓, poly(methyl methacrylate)
at 35, 55, 80 and 100 °C].

similar problem exists in a priori estimates of partial molar
volumes of supercritical components even in low molecular
weight liquids (16). The principle upon which Eq (3) is based
remains valid, however, and while the total amount of unrelaxed
volume may be available for a penetrant, the magnitude of C_H'
depends strongly on how condensible the penetrant is, since
this factor determines the relative efficiency with which the
component can utilize the available volume.

Transport. A companion transport model that also
acknowledges the fact that penetrant may execute diffusive
jumps into and out of the two sorption environments expresses
the local flux, N, at any point in the polymer in terms of a
two part contribution (17-20):

$$N = - D_D \frac{\partial C_D}{\partial x} - D_H \frac{\partial C_H}{\partial x} \tag{4}$$

where D_D and D_H refer to the mobility of the dissolved and
Langmuir sorbed components, respectively. It is typically
found that D_D is considerably larger than D_H except for non
condensible gases such as helium (14). The above expression
can be written in terms of Fick's law with an effective
diffusion coefficient, $D_{eff}(C)$, that is dependent on local
concentration:

$$N = - D_{eff}(C) \frac{\partial C}{\partial x} \tag{5}$$

The dual mobility model expresses the concentration dependency
of $D_{eff}(C)$ in terms of the local concentration of dissolved
penetrant, C_D, as shown in Eq (6):

$$D_{eff}(C) = D_D \left[\frac{1 + \dfrac{FK}{(1 + \alpha C_D)^2}}{1 + \dfrac{K}{(1 + \alpha C_D)^2}} \right] \tag{6}$$

where $F \equiv D_H/D_D$, $K \equiv C_H'b/k_D$ and $\alpha \equiv b/k_D$. This model explains
concentration dependency of the local diffusion coefficient
such as that shown for CO_2 in poly(ethylene terephthalate)
(PET) in Figure 6 (21) in terms of a progressive increase in
the fraction of the local concentration present in the higher
mobility Henry's law environment as the local Langmuir capacity
saturates at increasingly higher pressures.
    The points in Figure 6 were evaluated from the
phenomenological permeability and sorption concentration data

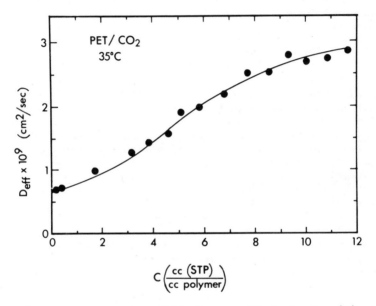

Figure 6. Local effective diffusion coefficient, $D_{eff}(C)$, for
carbon dioxide in poly(terephthalate) at 35 °C.

using a method that does not depend on the dual mode model in
any way (22). Interestingly, however, the line through the
data points corresponds to the predictions of $D_{eff}(C)$ using
Eq (6) along with the independently determined dual mode
parameters for this system (21,23). It is also important to
note that the form of data in this plot which exhibits a
tendency to asymptote at high pressures is not typical of
plasticization. Finally, if one considers the low concentration
region of Figure 6, it is clear that the diffusion coefficient
is surprisingly concentration dependent. For example, $D_{eff}(C)$
increases by more than 35% as the local sorbed concentration
rises from 0.142 cc(STP)/cc polymer (or 0.00020 wt. fraction) at
50 mm Hg to 1.7 cc(STP)/cc polymer (or 0.0025 wt. fraction) at
760 mm Hg. The above concentrations are extremely dilute,
corresponding to less than 1 CO_2/1200 PET repeat units and 1
CO_2/98 repeat units respectively (4).

An even more dramatic case corresponds to CO_2 in PVC in
which $D_{eff}(C)$ rises by 86% as the local concentration increases
from 1 CO_2/4400 repeat units at 100 mm Hg to 1 CO_2/1300 repeat
units at 500 mm Hg at 40°C (24). The above values of CO_2
solubility are based on measurements by Toi (24) and Tikhomorov
et al. (25) which are in good agreement.

At such extraordinarily low penetrant concentrations,
plasticization of the overall matrix is certainly not
anticipated. Motions involving relatively few repeat units are
believed to give rise to most short term glassy state
properties. In rubbery polymers, on the other hand, longer
chain concerted motions occur over relatively short time scales,
and one expects plasticization to be easier to induce in these
materials. Interestingly, no known transport studies in rubbers
have indicated plasticization at the low sorption levels noted
above for PVC and PET.

The discussion directly following Eq (6) provides a simple,
physically reasonable explanation for the preceding observations
of marked concentration dependence of $D_{eff}(C)$ at relatively low
concentrations. Clearly, at some point, the assumption of
concentration independence of D_D and D_H in Eq (6) will fail;
however, for our work with "conditioned" polymers at CO_2
pressures below 300 psi, such effects appear to be negligible.
Due to the concave shape of the sorption isotherm, even at a CO_2
pressure of 10 atm, there will still be less than one CO_2
molecule per twenty PET repeat units at 35°C. Stern (26) has
described a generalized form of the dual mode transport model
that permits handling situations in which non-constancy of D_D
and D_H manifest themselves. It is reasonable to assume that the
next generation of gas separation membrane polymers will be even
more resistant to plasticization than polysulfone, and cellulose
acetate, so the assumption of constancy of these transport
parameters will be even more firmly justified.

Although not necessary in terms of phenomenological
applications, it is interesting to consider possible molecular
meanings of the coefficients, D_D and D_H. If two penetrants
exist in a polymer in the two respective modes designated by "D"
and "H" to indicate the "dissolved" (Henry's law) and the "hole"
(Langmuir) environments, then the molecules can execute
diffusive movements within their respective modes or they may
execute intermode jumps.

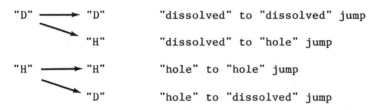

"D" ⟶ "D"	"dissolved" to "dissolved" jump
↘ "H"	"dissolved" to "hole" jump
"H" ⟶ "H"	"hole" to "hole" jump
↘ "D"	"hole" to "dissolved" jump

Clearly, the true character (activation energy, entropy and jump
length) of the phenomenologically observed D_D will be a weighted
average of the relative frequency of "D"→"D" and "D"→"H" jumps,
and likewise for D_H in terms of "H"→"H" and "H"→"D" jumps. Given
the relatively dilute overall volume fraction associated with
the non equilibrium gaps which comprise the "H" environment ($<$ 4
to 5% on a volume basis), one may to a first approximation
assume that most diffusive jumps of a penetrant from a "D"
environment result in movement to another "D" environment and
most diffusive jumps from "H" environments result in movement to
a "D" environment. The observed activation energies, entropies
and jump lengths, therefore, have fairly well-defined meanings
on a molecular scale.

One can easily show that the appropriate equation derived
from the dual mode sorption and transport models for the steady
state permeability of a pure component in a glassy polymer is
given by Eq (7) (18) when the downstream receiving pressure is
effectively zero and the upstream driving pressure is p.

$$P = k_D D_D \left[1 + \frac{FK}{1 + bp} \right] \tag{7}$$

The first term in Eq (7) describes transport related to the
Henry's law environment, while the second term is related to the
Langmuir environment. The tendency for the permeability to
asymptotically approach the limiting value of $k_D D_D$ at high
pressures derives from the fact that after saturation of the
upstream Langmuir capacity at high pressures, additional
pressure increases result in additional flux contributions only
from the term related to Henry's law which continues to increase
as upstream pressure increases.

The remarkable efficacy of the dual mode sorption and

transport model for description of pure component data has been illustrated by plots of the linearized forms of Eq (2) and Eq (7) for a wide number of polymer/penetrant systems (4,5,9, 10,22,24). These linearized plots are stringent tests of the ability of the proposed functional forms to describe the phenomenological data. Assink (27) has also investigated the dual mode model using a pulsed NMR technique and concluded that:

> "We have been able to demonstrate
> the basic validity of the
> assumptions on which the dual mode
> model is based and we have shown
> the usefulness of NMR relaxation
> techniques in the study of this
> model."

Whereas the dual sorption and transport model described above unifies independent dilatometric, sorption and transport experiments characterizing the glassy state, an alternate model offered recently by Raucher and Sefcik provides an empirical and fundamentally contradictory fit of sorption, diffusion and single component permeation data in terms of parameters with ambiguous physical meanings (28). The detailed exposition of the dual mode model and the demonstration of the physical significance and consistency of the various equilibrium and transport parameters in the model in the present paper provide a back drop for several brief comments presented in the Appendix regarding the model of Raucher and Sefcik.

Mixed Component Sorption and Transport

Arguments similar to those presented above for pure components have been extended to generalize the expressions given in Equations 2 and 7 to account for the case of mixed penetrants (29,30). The appropriate expressions are given below:

$$C_A = k_{DA}P_A + \frac{C'_{HA}b_A P_A}{1 + b_A P_A + b_B P_B} \qquad (8)$$

$$P_A = k_{DA}D_{DA} \left[1 + \frac{F_A K_A}{1 + b_A P_A + b_B P_B} \right] \qquad (9)$$

In the above expressions, "A" refers to the component of primary interest while "B" refers to a second "competing" component.

The permeability of a polymer to a penetrant depends on the multiplicative contribution of a solubility and a mobility term. These two factors may be functions of local penetrant concentration in the general case as indicated by the dual mode model. Robeson (31) has presented data for CO_2 permeation in

polycarbonate in which both solubility and diffusivity are
reduced due to antiplasticization caused by the presence of the
strongly interacting 4-4´ dichloro diphenyl sulfone. On the
other hand, sorption of a less strongly interacting penetrant,
such as a hydrocarbon, may affect primarily only the solubility
factor without significantly changing the inherent mobility of
the penetrant in either of the two modes. Flux reduction in
this latter context occurs simply because the concentration
driving force of penetrant A is reduced. This results from the
exclusion of A by component B from Langmuir sorption sites which
were previously available to penetrant A in the absence of
penetrant B.

Consistent with the preceding discussion concerning
sorption and flux reductions by relatively non interacting
penetrants, the data shown in Figure 7 clearly illustrate the
progressive exclusion of CO_2 from Langmuir sorption sites in
poly(methyl methacrylate) (PMMA) as ethylene partial pressure
(p_B) is increased in the presence of an essentially constant CO_2
partial pressure of p_A = 1.53±0.04 atm (32). The tendency of
the CO_2 sorption shown in Figure 7 to decrease monotonically
with ethylene pressure provides impressive support for the
"competition" concept on which Eq (8) and Eq (9) are based.
Permeation data are not available for this system to determine
if changes in the value of D_D and D_H occur in the mixed
penetrant situation; however, for the case of the relatively non
interacting ethylene at the pressures studied, any such effects
are expected to be minor.

The data shown in Figure 8 illustrate the reduction in
permeability of polycarbonate to CO_2 caused by competition
between isopentane and CO_2 for Langmuir sorption sites (33).
The flux depression shown in Figure 8 was found to be
reversible, with the permeability returning to the pure CO_2
level after sufficient evacuation of the isopentane-contaminated
membrane. Even at the low isopentane partial pressure
considered (117 mm Hg) the tendency is clear for the CO_2
permeability to be depressed from its pure component level
toward the the limiting value (P_A = $k_{DA}D_{DA}$) corresponding to
complete exclusion of CO_2 from the Langmuir environment. Further
increases in the isopentane partial pressure in the feed should
eventually complete the depression of CO_2 permeability to its
limiting value of 4.57 Barrers unless plasticizing effects set
in at the higher isopentane levels. This suggests, for example,
that the effective CO_2 permeability at a CO_2 partial pressure of
2 atm could be reduced by as much as 36% due to small amounts of
such hydrocarbons in the feed stream.

The lines drawn through the data in Figure 8 were
calculated from Eq (7) and Eq (9) for the pure and mixed
penetrant feed situations, respectively, using the same CO_2
model parameters in both cases. The affinity constant of

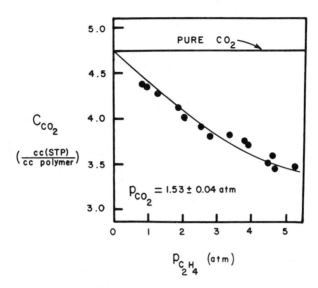

Figure 7. The effect of ethylene on the sorption level of carbon dioxide in poly(methyl methacrylate) at 35 °C.

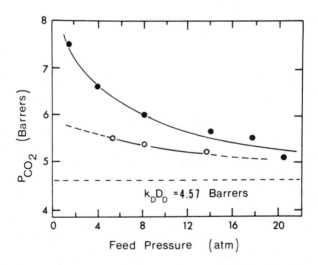

Figure 8. Permeability of Lexan polycarbonate to carbon dioxide at 35 °C as a function of CO_2 partial pressure. [○, in the presence of 117 mm. Hg of isopentane in the feed; ●, pure CO_2 .]

isopentane was estimated for use in Eq (9) to be 13.8 atm^{-1}
(33). The excellent fit of the data suggests that D_D and D_H for
$\overline{CO_2}$ in the mixed feed case are not affected measurably by the
presence of the relatively non-interacting isopentane. The
sorption and transport parameters for CO_2 in the polycarbonate
sample used in the above study are reported in Table 2.

Conclusions

The confluence of current views of the glassy state with
recent data for sorption and transport of mixed penetrants in
glassy polymers provides a complementary framework useful in
both areas. Specifically, the concept of "unrelaxed volume"
invoked to explain concavity in the sorption isotherm may
eventually be valuable in rationalizing different creep, impact
strength and physical aging behavior of glassy polymers. The
dual mode sorption and transport model provides a valuable basis
for both qualitative and quantitative analysis of gas separator
operation in the preplasticizing regime. Although plasticizing
effects are likely to be of considerable importance at high
pressures in gas separators with the present generation of
membrane materials, one might anticipate that the next
generation of materials based on very high T_g , highly rigid
backbone polymers will minimize sensitivity to such deleterious
effects. If such stiff-chain materials can be formulated into
useful membranes, phenomena associated with dual mode sorption
and transport will be of primary importance over the full range
of operating pressures, since the transition to plasticizing
behavior shown in Figure 1 would presumably require very high
sorbed concentration levels. Such plasticization-resistant
membranes might even maintain reasonable selectivity in
liquid/liquid permeation separations such as pervaporation.

Appendix: Comments Concerning The "Matrix" Model For
Sorption and Difffusion in Glassy Polymers

In the preceding discussion we have presented a model with
physically interpretable parameters to explain sorption data for
penetrants in glassy polymers. The basis of the model, the
concept of "unrelaxed volume" is also useful in understanding
many other properties of glassy polymers such as impact
strength, physical aging and creep, which are related to the non
equilibrium nature of these materials. Moreover, this model
yields a physically consistent expression for the local
effective diffusion coefficient, $D_{eff}(C)$, which shows
concentration dependence at sorbed concentrations well below the
point at which substantial interaction of one penetrant with the
matrix is generally expected to significantly facilitate the
movement of another penetrant.

Table 2

Sorption and transport parameters of CO_2 in
Lexan polycarbonate at 35°C

k_D $\dfrac{cc(STP)}{cc\text{-}atm}$	0.6751
b atm^{-1}	0.2563
C_H' $\dfrac{cc(STP)}{cc\ polymer}$	17.61
D_D cm^2/sec	5.1484×10^{-8}
D_H cm^2/sec	5.8266×10^{-9}

The dual mode model assumes that the heterogeneity that characterizes the glassy polymer state can be treated in a physically reasonable fashion using the concept of two basically different environments in the polymer. One environment corresponds to that in which a penetrant molecule is sorbed between chain segments that have achieved essentially their equilibrium conformations and interchain distances (perhaps with the aid of the invading penetrant during the "conditioning" step shown in Figure 3). A second environment corresponds to that in which a penetrant molecule finds itself sorbed in a region of localized lower density (perhaps due to local chain kinks or other impediments to rapid relaxation). The model requires, however, that the local concentrations of C_D and C_H in Eq (1) adjust themselves consistent with the requirement that there is only one chemical potential applicable to the penetrant, and at sorption equilibrium this chemical potential is also identical to that of the penetrant in the external gas phase. Any significant energetic heterogeneity associated with sorption in the nonequilibrium gaps regions should manifest itself as a failure in the Langmuir form to fit the deviation from the simple Henry's law model, since the Langmuir form assumes a uniform site affinity. Repeated tests of the linearized form of the Langmuir contribution to sorption have shown excellent conformity to the model. Any such energetic heterogenity is, therefore, of negligible importance with respect to sorption modeling. In their recent paper, Raucher and Sefcik interpret Dr. R. Assink's pulsed NMR study of ammmonia sorption in poly-styrene (27) and indicate that: "Spectroscopic analyses of gas molecules within polymer matrices are consistent with all of the gas molecules being in a single state" (28). One is given the indirect and erroneous impression from Raucher's and Sefcik's reference to Assink's work that the Assink study did not support the concepts of the dual mode model. On the contrary, the conclusions of Dr. Assink with regard to this issue completely support the dual mode model premises as shown below in a direct quote of the complete conclusions from the Assink paper (27):

"We have been able to critically examine the dual mode model by pulsed NMR relaxation techniques. The pressure dependence of the concentration of sorbed gas was consistent with the dual mode model while the relaxation data addressed itself to the validity of the assumptions made by the model. The assumption of rapid interchange was found to be valid for this system while the assumption of an immobile adsorbed phase could introduce a small error in the analysis. It should be possible to reduce this error by more exact measurements of the concentration of sorbed gas as classical pressure experiments could

provide. We have been able to demonstrate the
basic validity of the assumption on which the
dual mode model is based and we have shown the
usefulness of NMR relaxation techniques in the
study of this model."
The only "small error", suggested in Assink's statement
concerning the dual mode model's assumptions, deals with the
earlier approximation by Vieth and Sladek (34) that D_H was equal
to zero. The work of Assink was performed prior to the
formulation of the dual mobility model which eliminates this
approximation and accomodates values of $D_H > 0$ (17,18).

In the present symposium, Dr. Raucher argued that the
observed curvature in gas sorption isotherms does not arise from
a site saturation mechanism such as we have described in the
preceding discussion. The form of the sorption isotherm and
local concentration dependent diffusion coefficient proposed by
Raucher and Sefcik are given below:

$$C = \frac{\sigma_o p}{1 + \alpha C} \qquad (A-1)$$

$$D_{eff}(C) = D_o[1 + \beta C] \qquad (A-2)$$

where D_o was identified as the zero concentration limit of the
diffusion coefficient and β was identified as a
plasticization-related parameter which indicates the relative
sensitivity of the local diffusion coefficient, $D_{eff}(C)$ to the
presence of other penetrants in the vicinity. While no physical
interpetation was given to the sorption parameters, the σ_o
parameter is clearly equal to the limiting slope of the
concentration versus pressure isotherm. During a question and
answer period following his presentation, Dr. Raucher explicitly
ruled out an interpretation of the parameter α in terms of a
site saturation mechanism similar to the Langmuir model.

Drs. Raucher and Sefcik have stated: "The apparent
concentration dependence of the sorption capacity in the matrix
model results from a decrease in a Henry's law-like constant as
the polymer adjusts to the sorbate" (28). The extent of the
interpretation of α offered by these authors is that: "α is a
parameter indicating the magnitude of the change in solubility
arising from changes in the gas-polymer matrix". Specifically,
what is missing from their discussion is a statement as to what
the nature of the hypothetical polymer/penetrant interactions
are that cause the concavity in the isotherm. In other words,
what change is induced in the nature of the polymer in the
presence of the penetrant? The "model" comprised of Eq (A-1)
and (A-2) has a strictly empirical basis. It seems incumbent
upon these authors to explain how the change in the polymer
matrix makes it more difficult to insert another penetrant

[Eq (A-1)], but at the same time makes it easier for the
penetrant to move through the matrix [Eq (A-2)]. The dual mode
model offers the physical interpretation that concavity in the
isotherm arises from a site saturation mechanism related to
filling of unrelaxed gaps--a phenomenon which is not only easily
visualized and understood, but also has been tested explicitly
with independently obtained dilatometric data (See Figure 5).

In an attempt to justify the assumption of plasticization
put forth in their interpretation of β in Eq (A-2), Raucher and
Sefcik compare transport data and ^{13}C NMR data for the CO_2/PVC
system. This comparison has several questionable aspects. To
relate local molecular chain motions to the diffusion
coefficient of a penetrant, one should use the so-called local
effective coefficient, $D_{eff}(C)$, such as shown in Figure 5 rather
than an average or "apparent" diffusion coefficient as was
employed by these authors. $D_{eff}(C)$ describes the effects of the
local sorbed concentration on the ability of the average
penetrant to respond to a concentration or chemical potential
gradient in that region.

Raucher and Sefcik, on the other hand, base their
comparisons between NMR data and transport data on the so-called
"apparent" diffusion coefficient defined by:

$$D_a = \ell^2/6\Theta \qquad\qquad (A-3)$$

where Θ is the observed time lag and ℓ is the membrane
thickness. D_a does not have a simple meaning equivalent to a
true molecular mobility unless both the time lag and
permeability are independent of upstream pressure. Since this
situation is not typically observed in glassy polymers, one must
use $D_{eff}(C)$ for comparison with complementary techniques that
probe molecular motion.

The seriousness of this oversight is apparent in Sefcik and
Schaefer's analysis of Toi's transport data (24) in terms of
their NMR results (28). The value of the so-called "apparent"
diffusion coefficient calculated from Toi's time lag data
increases by ~25% for an upstream pressure range between 100 mm
Hg and 500 mm Hg. On the other hand, the value of $D_{eff}(C)$
calculated from Toi's data changes by 86% over the concentration
range from 100 to 500 mm Hg. The difference in the two above
coefficients arises from the fact that D_a is an average of
values corresponding to a range of concentrations from the
upstream value to the essentially zero concentration downstream
value in a time lag measurement. $D_{eff}(C)$, on the other hand,
has a well-defined point value at each specified concentration
and is typically evaluated (independent of any specific model
other than Fick's law) by differentation of solubility and
permeability data (22).

A second issue clouding the NMR interpretation of Toi's

$40°C$ transport data lies in the fact that the cited NMR data
were collected at $26°C$ rather than $40°C$. If one neglects this
fact, Sefcik and Schaefer suggest that the 5.7% increase in the
average rotating frame relaxation rate, $\langle R_{1\rho}(C) \rangle$ over the
pressure range (100 mm Hg to 500 mm Hg) corresponding to Toi's
measurements validate their claims concerning plasticization.
This suggestion neglects the fact that the 25% increase in D_a
does not match the 5.7% increase in $\langle R_{1\rho}(C) \rangle$. Moreover, recall
that D_a is actually not even the correct coefficient to compare
with $\langle R_{1\rho}(C) \rangle$. The observed 86% increase in $D_{eff}(C)$ is in poor
agreement with the 5.7% increase in the NMR parameter over the
range of pressures where both types of data are available.

The considerable discrepancy between the changes in the two
parameters $D_{eff}(C)$ and $\langle R_{1\rho}(C) \rangle$ is not impressive proof for
plasticiation at such low concentrations. This observation is
especially true since part of the observed 5.7% increase in
$\langle R_{1\rho}(C) \rangle$ may be accounted for by spin-spin effects. The value
of $\langle R_{1\rho}(C) \rangle$ requires cautious interpretation in terms of bulk
polymer properties such as the effective diffusion coefficient;
especially in light of the preceding discussion concerning Toi's
data. It is important to remember that even at 500 mm Hg, there
is only one CO_2 molecule for every 1300 PVC repeat units in the
above situation and yet an increase of 86% occurs in the value
of $D_{eff}(C)$ between 100 mm Hg and 500 mm Hg. These conditions are
extraordinarily dilute, and the observed concentration
dependence of $D_{eff}(C)$ is perfectly consistent with a site
saturation type behavior inherent in the dual mode model. We
have always acknowledged that concentration dependence of D_D and
D_H may become important factors at elevated pressures for
plasticization prone polymers. Such effects, if present have
been of second order importance in our work which is generally
performed on "conditioned" polymers at CO_2 pressures no greater
than 300 psi. Eventual concentration dependence of D_D and D_H
is, however, not related to the views of Raucher and Sefcik
summarized by Eq (A-1) and Eq (A-2). These equations require
that the seemingly mutually exclusive processes of
antiplasticization (Eq A-1) and plasticization (Eq A-2) occur
simultaneously.

Finally, it was suggested by Raucher and Sefcik that the
equations derived from their analysis were essentially as
effective as the dual mode model for describing existing data.
This statement requires some strong qualification, since no
linear equation can describe the inflected form of the $D_{eff}(C)$
plot shown in Figure 6 which was evaluated by graphical
differentiation of the permeability and solubility data without
reference to any particular model other than Fick's law (21,22).
The curve through the data, nevertheless, corresponds to the
dual mode model and provides a very good description of the
data. The form of these data is not at all consistent with

plasticization. This same argument will apply for essentially
all of the cases in which the dual mode model has been shown to
fit the form of the local diffusivity. The data points can at
best be made to "snake" around the linear fit of the expression
in Eq (A-2) and the linear model becomes seriously in error as
D_{eff} begins to asymptote. Limited data for two gas/polymer
systems were shown by Dr. Raucher which indicated that the
descriptions of permeabilities and time lags using the dual mode
and matrix models is not very different for the systems checked.
If this is true, it is surprising since the different forms of
$D_{eff}(C)$ would be expected to give rise to somewhat different
time lag results when integrated according to Frisch's method
(35). A similar conclusion pertains to the permeablity, since
even if the solubility data were perfectly described, the misfit
of $D_{eff}(C)$ suggests that there should be a misfit in
permeability as well. It is possible that a misfit of $D_{eff}(C)$
versus C and of C versus p could offset each other.

In one of their comparisons between the matrix model and
the dual mode model, a somewhat misleading presentation of data
is unintentionally offered in Figure 4 of the Raucher and Sefcik
paper: "Matrix Model of Gas Sorption and Diffusion in Glassy
Polymers" (28). What should be compared in the left hand side
of this plot is the dual mode model with refitted parameters
over the same pressure range as the "Matrix model" parameters
were refitted over. Clearly, a detailed statistical comparison
of the permeability and time lag predictions arising from Eq
(A-1) and Eq (A-2) must be made with the large body of
experimental data available for glassy systems before a
conclusion can be reached regarding the efficacy of the "matrix"
model for phenomenologically describing such data in general.

On a more basic level, since the matrix model implicitly
requires a somewhat inconsistent interpretation for the various
model parameters in Eq (A-1) and Eq (A-2), it becomes primarily
an empirical means of reproducing the observed pure component
data with no fundamental basis for generalization to mixtures.
One could, of course envision several extensions based on
additional α terms in the denominator of Eq (A-1) and additional
β terms in Eq (A-2). Such an approach to mixture permeation
analyses would be completely empirical and mimic the
generalization of Eq (2) and Eq (7); however, without any
physical justification. The generalizations of Eq (2) and Eq
(7) were natural outgrowths of the fundamental physical basis of
the Langmuir isotherm. The fact that the mixture data are so
consistent with Eq (7) and Eq (9) provides strong support for
the physical basis of the dual mode model.

An important value of a permeation model is not simply its
ability to correlate experimental data, but rather to provide a
framework for understanding the principal factors controlling
membrane performance. The dual mode model is derived from

consistent, physical descriptions of the glassy state and yields parameters which make unambiguous molecular scale statements regarding the general nature of glassy, amorphous materials.

In summary, the two expressions offered in Eq (A-1) and Eq (A-2) provide empirical forms for correlating the general trends in pure component sorption and transport data. These expressions are not offered in terms of internally consistent physical arguments and appear to offer no fundamental basis for understanding general glassy state behavior. Furthermore, no fundamental approach is apparent for the treatment of the critically important mixed penetrant problem using the model, since the various parameters in the model lack the well-defined significance provided by the dual mode model.

Nomenclature

b_i Affinity constant of component i for the polymer (atm^{-1})
D_D Diffusion coefficient of the Henry's law species (cm^2/sec)
D_H Diffusion coefficient of the Langmuir species (cm^2/sec)
F Ratio of Langmuir and Henry's law diffusion coefficients
k_D Henry's law constant (cm^3 gas (STP)/cm^3 polymer-atm)
K $C'_H b/k_D$ where C'_H is the capacity of Langmuir mode (cm^3 gas (STP)/cm^3 polymer)
p_i Upstream driving pressure for penetrant i (atm)
P_i Permeability of i (cm^3 gas(STP)-cm/cm^2-sec-cmHg)
V_g Specific volume of the glassy polymer (cm^3/g)
V_ℓ Specific volume of the densified glassy polymer (cm^3/g)

Acknowledgments

The authors gratefully acknowledge support of this work under NSF Grant No. CPE-79-18200 and ARO Contract No. DAAG29-81-K-0039. Also the assistance of Mrs. Leslie Edgerton in typing the manuscript is acknowledged.

Literature Cited

1. Stannett, V. T.; Koros, W. J.; Paul, D. R.; Lonsdale, H. K. and Baker, R. W.; Adv. in Polymer Sci., 1979, 32, 69.
2. Barrer, R. M. and Barrie, J. A.; J. Polym. Sci., 1957, 23, 331.
3. Pye, D. G.; Hoehn, H. H. and Panar, M.; J. Appl. Polym. Sci., 1976, 20, 287.
4. Koros, W. J. and Paul, D. R.; J. Polym. Sci.-Polym. Phys. Ed., 1978, 16, 1947.
5. Wonders, A. G. and Paul, D. R.; J. Membr. Sci., 1979, 5, 63.
6. Fechter, J. M.; Hopfenberg, H. B. and Koros, W. J.; Polym. Engr. and Sci., 1981, 21, 925.

7. Koros, W. J. and Paul, D. R.; J. Polym. Sci., Polym. Phys.
    Ed., 1981, 19, 1655.
8. Horiuti, J.; Sci. Papers of the Inst. of Phy. and Chem.
    Res. (Tokyo); 1931, 17, 126.
9. Chan, A. H. and Paul, D. R., Polym. Engr. and Sci., 1980,
    20, 87.
10. Huvard, G. S.; Stannett, V. T.; Koros, W. J. and
    Hopfenberg, H. B., J. Membr. Sci., 1980, 6, 185.
11. Chen, S. H., M.S. Thesis, North Carolina State University,
    1982.
12. Lewis, O. G.; "Physical Constants of Linear Homopolymers,"
    Springer Verlag, New York (1968).
13. Kolb, H. J. and Izard, E.; J. Appl. Phys., 1949, 20, 564.
14. Matuska, S. and Ishida, Y.; J. Polym. Sci., Pt. C., 1960,
    14, 247.
15. Mercier, J. P.; Aklonis, J. J.; Litt, M.; and Tobolsky, A.
    V.; J. Appl. Polym. Sci., 1965, 9, 447.
16. Sandler, Stanley I.; "Chemical and Engineering
    Thermodynamics," J. Wiley and Sons, New York, 1977, pg.
    435.
17. Petropoulos, J. H.; J. Polym. Sci., Pt. 2-A, 1970, 8,
    1797.
18. Paul, D. R. and Koros, W. J.; J. Polym. Sci.-Polym. Phys.
    Ed., 1976, 14, 675.
19. Koros, W. J.; Chan, A. H. and Paul, D. R.; J. Membr. Sci.,
    1977, 2 165.
20. Barrie, J. A.; Williams, M. L. and Munday, K.; Polym. Engr.
    and Sci., 1980, 20, 20.
21. Koros, W. J., Ph.D. Dissertation; "Sorption and Transport
    of Gases in Glassy Polymers," The University of Texas
    (Austin), 1977.
22. Koros, W. J.; Paul, D. R. and Rocha, A. A.; J. Polym. Sci.,
    Polym. Phys. Ed., 1976, 14, 687.
23. Koros, W. J. and Paul, D. R.; J. Poly. Sci., Polym. Phys.
    Ed., 1978, 16, 2171.
24. Toi, K.; Polym. Engr. and Sci., 1980, 20, 30.
25. Tikhomorov, B. P.; Hopfenberg, H. B.; Stannett, V. T. and
    Williams, J. L.; Macromol. Chem., 1968, 118, 177.
26. Stern, S. A. and Saxena, V.; J. Membr. Sci., 1980, 7, 47.
27. Assink, R. A.; J. Polym. Sci.-Polym. Phys. Ed., 1975, 13,
    1665.
28. Raucher, D. and Sefcik, M.; Paper presented in this
    symposium.
29. Koros, W. J.; J. Polym. Sci.-Polym. Phys. Ed., 1980, 18,
    981.
30. Koros, W. J.; Chern, R. T.; Stannett, V. T. and Hopfenberg,
    H. B.; J. Polym. Sci.-Polym. Phys. Ed., 1981, 19, 1513.
31. Robeson, L. M.; Polym. Engr. and Sci., 1969, 9, 277.

32. Sanders, E. S.; Koros, W. J.; Hopfenberg, H. B. and
    Stannett, V. T.; "Mixed Gas Sorption in Glassy Polymers:
    Equipment Design and Considerations and Preliminary
    Results," Submitted for Publication, J. Polym. Sci.- Phys.
    Ed.
33. Chern, R. T.; Koros, W. J.; Hopfenberg, H. B. and Stannett,
    V. T.; "The Effects of Low Partial Pressures of Isopentane
    on the Permeability of Polycarbonate to CO_2," accepted by
    J. Polym. Sci.- Phys. Ed.
34. Vieth, W. R. and Sladek, K. J.; J. Coll. Sci., 1965, 20,
    1014.
35. Frisch, H. L.; J. Phys. Chem., 1957, 61, 93.
36. Smith, G. N.; M.S. Thesis, North Carolina State University,
    1980.

RECEIVED December 27, 1982

Standard Reference Materials for Gas Transmission Measurements

JOHN D. BARNES

National Bureau of Standards, Polymer Science and Standards Division,
Washington, DC 20234

Standard Reference Material 1470 is a 23 micrometer thick polyester film whose gas transmission
characteristics with respect to helium, carbon dioxide,
oxygen, and nitrogen have been carefully measured. A
completely computerized manometric permeation measuring facility developed at NBS was used for the measurements. The steps taken to characterize the gas transmission rate of this material over the range of pressures from 67.5 kPa to 135 kPa and over the range of
temperatures from 18 °C to 31 °C are described. The
results obtained in these measurements are compared with
those in the literature. The role of Standard Reference
Material 1470 in improving the repeatability and reproducibility of gas transmission measurements employing
other instrumentation is discussed.

A Standard Reference Material (SRM) can be defined as "a material (not necessarily a pure substance) having given properties
with numerically assessed values, within certain tolerances, certified by an appropriate technical body" (1). An SRM for oxygen
gas transmission measurements has been available since 1978 from
the National Bureau of Standards (NBS) (2). This material has recently been certified for its permeance with respect to carbon
dioxide, helium, and nitrogen as well as oxygen. This was done to
make the material available to a larger community of users. The
work on oxygen was repeated in order to determine whether the
existing stock of material had aged to any appreciable extent and
to ensure that all results were reported on a common basis. The
present paper describes the experiments that were done to characterize the gas transmission properties of the material over a
range of conditions that the user can expect to experience and the
manner in which the results on the SRM certificate are to be interpreted by the user.

SRM's exist because of the widely perceived need to achieve
agreement among measurement results obtained from a measurement
system. In order for such materials to be useful, the property of
interest must be stable in the chosen material, the property must
be measured by a method that is free of bias, and the property
must be homogeneous across all possible specimens of the material.
    The typical user of an SRM performs a measurement of the
certified property using the apparatus in his laboratory. He
compares his result with the certificate value and, if his measurement fails to reproduce the certified value within acceptable
limits of precision he investigates his measurement system to remove sources of bias. There are numerous examples of measurement
systems in which the level of agreement among measured results by
employing SRM's in this way, and the statistical assessment of the
performance of measurement systems employing SRM's is a well-
established science (3).
    NBS's involvment in SRM's for gas transmission studies arose
out of activities within ASTM Committees D-20 (Plastics) and F-2
(Flexible Barrier Materials) that were aimed at standardizing a
new instrumental method employing a coulometric oxygen detector
for the measurement of oxygen gas transmission rates in materials
used for various kinds of packaging (4). A subsequent reevalua-
tion of the precision data for the classical manometric and volumetric methods (5) for gas transmission measurements showed that
there are serious problems with the reproducibility and repeatability these methods. The use of SRM 1470 to verify the calibration
is currently being written into ASTM standards D-3985 and D-1434.

Approach

    The effort being described here is a simple extension of the
work that was done earlier to characterize the gas transmission
properties of SRM 1470 with respect to oxygen alone (2). The
material that is used for this SRM is a poly(ethylene terephthalate) (PET) film approximately 23 micrometers thick that is used
commercially as a base for magnetic tape. This material appeared
to be suitable from the standpoint of homogeneity and long term
stability and its oxygen transmission rate was in the desired
range. Similar material was already in use as a de facto SRM provided by a manufacturer of coulometric oxygen transmission measuring equipment. Specimens for the recertification measurements
were obtained by random sampling from the existing stock of
material.
    We chose to use a modernized version of the manometric
measuring technique both in our earlier work and in the present
experiments. This approach offered good flexibility in that the
electronic manometers used to monitor the gas flux respond to a
wide variety of gases. It is also possible to calibrate the
apparatus reliably using existing standards. By completely
computerizing the measurement process (6) we were able to provide

for unattended operation as well as enhanced reliability and
safety.

In the time-lag technique (7) for measuring permeation gas
passes through a well defined area of a film from a reservoir at
an elevated pressure into a receiving chamber. The film is devoid
of permeant to begin with. In the manometric realization of this
method the accumulation of permeant in the receiving chamber is
measured by monitoring the pressure of the gas in the chamber. In
our system we begin by connecting both chambers to vacuum pumps to
remove any sorbed gases. We then monitor the small pressure rise
that occurs when the receiving chamber is isolated from the pump
but the upstream chamber is still being evacuated. We assign this
pressure rise to a background flux. At a chosen instant the upstream chamber is supplied with gas at a controlled pressure. Gas
does not appear in the downstream chamber until enough of it diffuses into the film to establish a finite concentration gradient
at the downstream boundary. After a while the pressure in the
downstream chamber rises at a steady rate. Once we are satisfied
that we have enough data to precisely estimate the steady-state
flux we terminate the pressure monitoring and reevacuate both
sides of the film in preparation for the next experiment.

Calibration

The calibration problem for our measurement system can be
discussed in the light of the following relationship:

$$n = pV/RT = AF_\infty [t-\tau \cdot q(t/\tau)] \tag{1}$$

where n is the number of moles of permeant accumulated in the
receiving chamber after a time t, p is the pressure in the receiving chamber at time t, V is the volume (in liters) of the receiving chamber, T is the temperature of the system (in kelvins),
R is the ideal gas constant (8.3144 kPa-1/mol-K) and A is the area
(in m^2) of film through which permeation takes place. F_∞ is the
permeant flux (in $mol/m^2 \cdot s$) for steady-state permeation, τ is
the time-lag and $q(t/\tau)$ is a function that accounts for the
transient response of the system. F_∞ is the product of the upstream gas pressure, p_u, and the permeance, P, of the film being
tested. The functional form of $q(t/\tau)$ for the case of ideal diffusion can be obtained by rearranging the equations from the standard texts (8,9). In our case the receiving volume, V, is large
enough and the final pressure is low enough so that we can neglect
the effects due to the finite volume of the receiving chamber.
Under these conditions there is a substantial period of time during which q is essentially equal to unity and the pressure vs
time trace may be treated as a straight line with slope s = F_∞
ART/V and intercept y0 = $-\tau AF_\infty RT/V$, a circumstance that our computer programs make use of to perform data reduction by least
squares.

The calibration characteristics of the pressure transducer are
described by the following relationship:

$$P_{true} = P_{app} \times [1 + a(T)] \qquad (2)$$

The subscript "true" denotes the corrected pressure values while
"app" refers to the observed values after they have been corrected
for contributions from the background flux. The function $a(T)$ is
a small correction that is obtained by intercomparing each cell
manometer with a well-calibrated manometer that is always evacuated and maintained at a fixed temperature. Differentiating equation 2 gives

$$s = [1 + a(T)]dp_{app}/dt = [1 + a(T)] \, s_{app} \qquad (3)$$

where s_{app} is the slope derived from fitting a straight line to
the apparent pressure vs time trace.
        In extracting s_{app} from our experimental data we use only
data for long times ($t > 2\tau$) and we use an iterative process to
correct for the small contribution from the time-lag.
        Using equations 1, 2, and 3 together with the definition of
the permeance we obtain

$$P = V \, [1 + a(T)] \, s_{app}/ART \, p_u \qquad (4)$$

as an expression linking measured quantities with the quantity we
wish to determine. The quantity $V[1 + a(T)]/ART$ can be regarded
as a temperature dependent "cell constant" whose value is fixed by
the design of the apparatus. Our studies indicate that this cell
constant is accurate to within 2 percent of its measured value and
that there is a random error of approximately 1 percent attribu-
able to the calibration instability of the pressure sensors. More
detailed studies of the calibration errors are currently in progress.
        Potential sources of interference that can bias the observed
results include leakage into the cell from extraneous sources and
outgassing of the film being tested. A guard ring is built into
the cell to minimize the former problem and we correct for the
latter effect by monitoring the pressure rise in the cell while
the upstream chamber is under a vacuum. If the resulting leakage
flux is too large the computer postpones the experiment until the
flux drops to an acceptable level. We perform the carbon dioxide
measurements separately precisely because the outgassing in this
case is so protracted as to interfere with the other measurements.
        If the leakage flux falls within the acceptable range we
merely measure its value for use as a baseline correction to the
pressure vs time data taken when the upstream chamber is pressur-
ized. The presence of this background flux tends to degrade the
signal to noise ratio of the data for the less permeable species.
Experiments using a solid aluminum plate in place of the film
demonstrate that this leakage flux arises mainly from the polymer.

Experimental Design

Nine sheets of material were selected at random from the
entire batch of SRM 1470 that was on hand. Our apparatus subjects
three specimens to the same treatment at the same time, where a
"treatment" is a measurement using a specified combination of gas,
temperature, and upstream pressure. Measurements were made at 5
different temperatures ranging from 18 °C to 31 °C in order to
allow the user some flexibility in his choice of measurement temperature. Measurements were also made at 4 different pressures
ranging from 67.5 kPa to 135 kPa in order to determine if the
material behaved non-ideally. Since we used a complete factorial
design the total number of treatments was 80 (4 gases x 5 temperatures x 4 pressures). Each specimen was subjected to the complete
set of treatments before being replaced by a new specimen. All
measurements involving nitrogen, helium, and oxygen at a given
temperature were made before the temperature was changed. The
treatments involving carbon dioxide were applied last because of
the prolonged outgassing required between experiments using this
gas. Except for the restrictions described above the treatments
were applied in random order. We found no evidence that the previous sample history affected the measurement results as long as
as care was taken to ensure proper outgassing before applying the
next treatment.

The order in which the treatments were to be applied was
programmed into the computer, and the apparatus then operated
unattended until it was time to insert a fresh data recording disc
disc or to change the specimens. A complete series of treatments
on a group of three specimens required about 6 weeks.

The processing of the data consisted of extracting the
steady state slopes and the time lags from the pressure versus
time traces, converting the slopes to permeances using the equations given above, and tabulating the results for subsequent statistical analysis.

Results

The complete set of experiments described above produced
approximately 720 values of the permeance and 540 values of the
time-lag (the time-lag for helium was too short to measure). We
used the following fixed effects statistical model to study the
data from each of the gases:

$$\ln Q = \mu_a + S_i + \theta_j + \rho_k + \varepsilon_{ijk} \qquad (6)$$

Q represents either the time-lag or the permeance as required,
the S_i term accounts for the effect of selecting a particular
one of the nine specimens, the θ_j term accounts for the effect

of choosing a particular one of the five temperatures, and ρ_k accounts for the pressure effect. The pressure effect was significant only when dealing with the time-lag data for carbon dioxide (see discussion below). We did not find any significant interaction terms between the main effects shown in equation 6. The ε_{ijk} term accounts for random errors of measurement. Over the narrow (13 °C) range of temperatures in our experiments the temperature effect is linear and a temperature coefficient can be derived by fitting the θ values to the following model:

$$\theta_j = \alpha + \beta \ (T_j - 23) \qquad (7)$$

The value of 23 °C was chosen so that the certificate values would be correct for the standard laboratory atmosphere (10).

Using equation 7 and taking the exponential function of both sides of equation 6 yields:

$$\hat{Q}(T) = \hat{Q}(23) \ e^{\beta(T-23)} \qquad (8)$$

where the "^" denotes that the quantity is an estimated value. Note that $Q(23)=e^{\mu}Q + \alpha$. A measurement on an individual specimen can be expected to deviate from this estimate because of measurement error and because of the specimen effect embodied in the S_i. Values of Q(23) and β obtained from the analysis of the permeance and the time-lag data are given in Tables I and II respectively. Since the time-lag for carbon dioxide depends upon the upstream pressure it is necessary to multiply the estimates obtained from equation 8 by a term of the form:

$$e^{a_1(p_u - 101.32) + a_2(p_u - 101.32)^2} \qquad (9)$$

where the upstream pressure is expressed in kPa. Estimates derived from equations 8 and 9 are valid only for pressures between 67.5 and 135 kPa and for temperatures between 18 °C and 31 °C.

Tables I and II provide additional statistical data that can be used to qualify the estimates derived from the fitting process. C_μ is the standard deviation of μ, its numerical value is largely determined by the sampling error arising from the selection of test specimens. C_s is the standard deviation of the S_i's, which is a measure of the inhomogeneity of the lot of SRM material. C_r is the standard deviation of the residuals from the fit, which is a measure of the extent to which individual data values depart from the model in equation 6. We have chosen not to construct the usual confidence or tolerance intervals because we do not have enough data on the distribution of the S_i's.

The individual values of the S_i are plotted as a function of gas in Figures 1 and 2. If the factor that is giving rise to this variability influences all gases equally we would expect the lines in Figures 1 and 2 to be parallel. Parallel straight lines

Table I   -   Summary of fit results for Permeance data

Gas

Parameter	N2	O2	CO2	He	Units
$\hat{P}(23)$	.042	.35	1.72	13.8	$pmol/m^2.s.Pa$
β	.0521	.0376	.0309	.0287	1/K
C_μ	.023	.014	.018	.012	
C_s	.057	.045	.056	.037	
C_r	.050	.017	.010	.009	

Table II   -   Summary of fit results for Time-lag data

Gas

Parameter	O2	N2	CO2	Units
$\tau(23)$	388.	1792.	2160.	sec.
β	-.0597	-.0763	-.0624	1/K
C_μ	.013	.022	.014	
C_s	.039	.067	.042	
C_r	.024	.094	.016	
a_1	--	--	-1.873×10^{-3}	1/kPa.
a_2	--	--	09.11×10^{-6}	$1/(kPa.)^2$

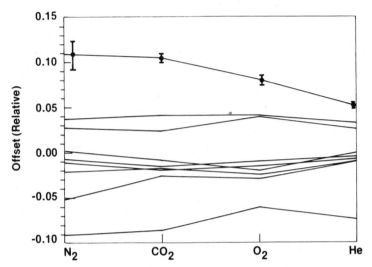

Figure 1.   Specimen effect for the permeance plotted as a function of gas.

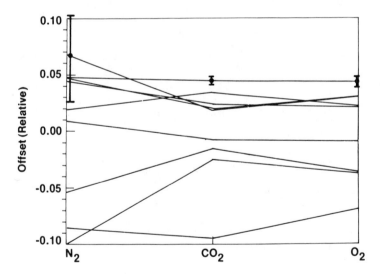

Figure 2.   Specimen effect for the time lag plotted as a function of gas.

account for most of the variation but the scatter shown in the figures is significant. We are studying additional specimens to better define the structure and distribution of these specimen effects.

We are also seeking simple auxiliary measurements such as the specimen thickness or the density that can be used to define correction factors for reducing the measured values to a common basis. In the absence of such a scheme the specimen-to-specimen variability is the dominant factor limiting the precision that can be expected when comparing measurements among different stations in a measurement system.

In some instances where SRM's have been found to be inhomogeneous it is possible to individually certify each specimen. Aside from the fact that such a procedure would lead to intolerably costly standards, it is impractical because we have found that specimens that have been removed from the measuring apparatus cannot be reused. Thus a unit of SRM 1470 contains about 15 sheets of material, which last the average user quite a long time.

The observed decrease of the ρ values for the time-lag of carbon dioxide with increasing pressure is consistent with published data that have been interpreted using a "partial immobilization" model (11,12). It is somewhat surprising that we do not observe a concomitant decrease in the permeance; perhaps the pressures used are too low or the pressure range is not broad enough.

The residuals from the fits to the short time part of the pressure versus time traces provide additional evidence that the transport process of carbon dioxide in SRM 1470 is different from that of the other gases. The trace for carbon dioxide departs from the background trend line much more slowly than do the traces for the other gases. This is qualitatively reasonable in terms of the partial immobilization model, but we hope to develop appropriate solutions to the transport equations to verify this effect.

The effects of partial immobilization are small in the range of conditions covered in our experiments. Users who wish to carry out measurements at much higher pressures and temperatures should realize that they may have to take explicit account of this phenomenon.

We are modifying our apparatus to operate at higher pressures and at temperatures encompassing the glass transition of PET.

Assuming a specimen thickness of 23 micrometers we calculated permeabilities and diffusion coefficients for comparison with literature values. Our data and the data used for comparison are listed in Table III, which is organized along the same lines as Yasuda's compilation (13). Our values compare well with the limited data available (14-18) when one makes allowance for the differences in film morphology that are likely to exist.

The gas transmission properties of PET films similar to SRM 1470 are near the low end of the range commonly associated with

Table III - Literature values of Transport Parameters

Gas	Ref.	$\underline{P}(25)$	E_P	$D(25)$	E_D	Comment
N2		1.074	38.0	5.73	55.6	SRM 1470
	14	2.01				
	17	2.18	32.7	14.	44.0	crystalline
	15	3.23				
	17	4.35	26.4	20.	47.7	amorphous
02	4	7.7				
	18	8.6				ASTM D-3985
		8.73	27.4	25.6	43.5	SRM 1470
	18	8.8				ASTM D-1434
	17	11.7	32.3	35.	46.1	crystalline
	15	12.6				
	17	19.7	37.6	50.	48.5	amorphous
CO2	18	22.7				ASTM D-1434
	16	39.4				
	12	40.6	24.8	3.5	49.7	
		42.1	22.6	4.62	45.5	SRM 1470
	17	56.9	18.4	6.	50.2	crystalline
	17	100.4	27.6	8.5	52.3	amorphous
HE	16	324.				
	14	335.				
		336.	20.9			SRM 1470
	17	442.	19.7	20000	20.	crystalline
	17	1098.	21.3	30000	19.3	amorphous

Explanation of selected column headings:

Ref.   - See Literature cited for source of data

$\underline{P}(25)$ - permeability at 25 °C expressed in amol/m . s. Pa   (1 amol = 10^{-18} mole)

E_P - Activation energy for Permeation, kJ/mol

$\underline{D}(25)$ - Diffusion coefficient as calculated from $D = \ell^2/6\tau$ expressed in $10^{-14} m^2/s (10^{-10} cm^2/s)$

E_D - Activation energy for Diffusion, kJ/mol

homopolymer films (13,19). Persons interested in good gas barriers would like to see SRM's of lower transmission rate. In other cases (notably including separation membranes) one desires a poor barrier to certain gases. NBS is working to identify needs for new standard materials and improved measurement methods in these areas.

Since it is difficult to control processes for making plastic films in order to limit variability in their physical properties to one percent or less we find that material inhomogeneity is an obstacle to the development of better SRM's. It is also difficult to use such materials as transfer standards because their characteristics often drift with time and test specimens often do not tolerate abuse suffered during the measuring process.

Needs for improved measurement methods differ depending on whether one is considering low or high transmission rate materials. In the former case one needs very sensitive detectors. Selectivity is also desirable so that interferences from extraneous species can be avoided. In the case of high transmission rate materials instrumental time constants and saturation effects need to be better understood. In all cases there is a need for more convenient instruments and a better knowledge of their operating principles.

The existence of biases between labs is a clear indication that the operating principles that are used to describe a measurement process are not being realized in practice. In some cases the operator's technique fails to account for a factor that influences the measurement. In other cases the model that is used to describe the measurement process contains approximations that are violated in using the method. Examples include the back pressure correction, which can be important in situations where the gas is highly soluble in the polymer, the correction for the volume increase in manometric measurements using mercury columns 5, 20, or the influence of humidity on permeability in systems using moist test gases or carrier gases (4,21). If SRM 1470 is exposed to a carrier gas near 100 percent humidity its transmission rate decreases by about 20 percent (21). For this reason ASTM D-3985 specifies that calibration and referee measurements must be made with dry gases (4) even though the coulometric apparatus normally uses moist gases. Biases in measurement systems often depend upon the level of the property being measured. It is, therefore, desirable to cover a broad range of levels when checking calibration.

The main purpose of our work is to provide a material that is easily available to anyone who wishes to evaluate the performance of a permeation measuring system. Users whose equipment can accept any of the gases we used can cover a 350-fold range of transmission rates with the values on the SRM certificate. Interlaboratory comparisons that have been conducted to date (22) reveal that the repeatability (level of agreement for results obtained within individual laboratories) is much better than the reproducibility (level of agreement between laboratories). This situation is common in measurement systems that have been operating without benefit of standards.

The resolution of an experiment to detect differences between laboratories will depend upon the statistical design that is used. Given the specimen to specimen variability that exists in SRM 1470, differences of less than 5 percent of the measured value will be difficult to detect using this material. Since differences between labs of 30 percent are common in the existing measurement system, there is considerable room for improvement. Individual laboratories may be able to achieve good agreement by regularly exchanging check samples of materials other than an SRM, but their results could well contain systematic errors if they fail to check their measurements against others in which bias has been carefully eliminated.

Literature Cited

1. Milazzo, G. "Standard Reference Materials and Meaningful Measurements", National Bureau of Standards Special Publication 408; Seward, R., Ed.; U. S. Govt. Printing Office, Washington, 1975, p. 127.
2. Barnes, J.D.; Martin, G.M.;" Standard Reference Materials: SRM 1470: Polyester Film for Oxygen Gas Transmission Measurements", National Bureau of Standards Special Publication 260-58; U.S. Govt. Printing Office, Washington, 1979.
3. Mandel, J. "Standard Reference Materials and Meaningful Measurements", National Bureau of Standards Special Publication 408; Seward, R., Ed.; U. S. Govt. Printing Office, Washington, 1975, P. 146.
4. ASTM Standard D-3985, "Standard Test Method for Oxygen Gas Transmission Rate through Plastic Film and Sheeting Using a Coulometric Sensor"; in Part 35 of Annual Book of ASTM Standards, American Society for Testing and Materials, Philadelphia, Revised Annually.
5. ASTM Standard D-1434, "Standard Test Methods for Gas Transmission of Plastic Film and Sheeting"; in Part 35 of Annual Book of ASTM Standards, American Society for Testing and Materials, Philadelphia, Revised Annually.
6. Barnes, J. D., Technical Papers Volume XXVIII, Society of Plastics Engineers, Inc.; Fairfield, CT; 1982, p. 19.
7. Daynes, H. A. Proc. Phil. Soc. London, 1920, 97A, 286-307.
8. Crank, J.; Park, G.S.; "Diffusion in Polymers"; Crank, J. and Park, G. S., Eds.; Academic Press; New York, 1968, p. 6.
9. Crank, J.; "Mathematics of Diffusion", 2nd ed.; Clarendon Press, Oxford, 1975 p. 49.
10. ASTM Standard E-171, "Standard Specification for Standard Atmospheres for Conditioning and Testing Materials"; in Part 35 of Annual Book of ASTM Standards, American Society for Testing and Materials, Philadelphia, Revised Annually.

11. Koros, W.J.; Paul, D.R.; Fujii, M.; Hopfenberg, H.B.;
     Stannet, V. - J. Appl. Polymer. Sci., 1977, 21, 2899-904.
12. Koros, W.J., and Paul, D.R. - J. Polym. Sci. - 1978, 16, 217-
     87.
13. Yasuda, H.; Stannett, V.; "Polymer Handbook"; J. Brandrup
     and E.H. Immergut, Eds.; Wiley- Interscience, New York,
     1975 pp. III-299-240.
14. Stern, S. A. "Membrane Processes for Industry, Proceedings
     of the Symposium", Southern Res. Inst., 1966, p. 196.
15. Tajar, T. G.; Miller, I. F. Amer. Inst. Chem. Eng. J.,
     1972, 18, p. 78.
16. Brubaker, D. W.; Kammermeyer, K. Ind. Eng. Chem., 45, 1148
     (1953).
17. Michaels, A. S.; Vieth, W. R.; Barrie, J. A. J. Appl. Phys.
     34(1), 1963, p. 13.
18. Fenelon, P. J.;"Polymer Science and Technology, Vol 6;
     Permeability of Plastics Films and Coatings to Gases,
     Vapors, and Liquids"; H. B. Hopfenberg, Ed.; Plenum Press,
     New York, pp. 285-299.
19. Hwang, S.T.; Choi, C. K.; Kammermeyer, K. - Separation
     Sci., 9(6), p. 461-78.
20. Evans, R. E. J. Testing and Evaluation, 1974, 2, 529-32.
21. Brown, C. N.; personal communication
22. Pike, L.; personal communication
23. "The International System of Units (SI)", National Bureau
     of Standards Special Publication 330; U.S. Govt. Printing
     Office, Washington, 1977.

Appendix   - A Word About Units

The system of units used to express our results is completely in accordance with the prescriptions of the SI system (23).
This scheme differs from the various systems in use in industry
and academia in that it uses the mole instead of the cc(STP) to
express the quantity of matter being transported, the pascal
rather than the atmosphere or the cm. Hg. to express pressure, the
meter rather than the mil, the inch, or the centimeter to express
length, and the second rather than the day to express time. Our
experience indicates that the existing variety of unit systems
leads to confusion and that calculations of related physical
properties such as permeabilities, diffusion coefficients, and
solubilities are easier using the SI units. More modern measurement systems which detect permeants by means of the electrical
currents generated by individual atoms are easier to analyze when
one uses moles rather than cc(STP) to express the amount of matter
undergoing transport. Applications involving the transport of
mixed permeant species are also easier to deal with on a molar
basis. Conversion tables between the SI units and customary units
are provided on the SRM certificate and in the appropriate standards documents (4,5).

We have chosen to express our results as permeances rather
than permeabilities because we are presently unable to justify
the assumption that the properties of the material are uniform
throughout the film. We suspect that normalizing the results to
unit thickness would introduce additional scatter. The SRM
certificate gives instructions for calculating gas transmission
rates, which are the quantities that are measured directly.

RECEIVED January 12, 1983

Gas Transport and Cooperative Main-Chain Motions in Glassy Polymers

DANIEL RAUCHER and MICHAEL D. SEFCIK
Monsanto Company, St. Louis, MO 63167

Carbon-13 rotating-frame relaxation rate measurements are used to elucidate the mechanism of gas transport in glassy polymers. The nmr relaxation measurements show that antiplasticization-plasticization of a glassy polymer by a low molecular weight additive effects the cooperative main-chain motions of the polymer. The correlation of the diffusion coefficients of gases with the main-chain motions in the polymer-additive blends shows that the diffusion of gases in polymers is controlled by the cooperative motions, thus providing experimental verification of the molecular theory of diffusion. Carbon-13 nmr relaxation measurements also show that the presence of a permanent gas alters the cooperative motions of the polymeric chains. These changes in main-chain molecular motions correlate with changes in the diffusion coefficient of the gas in the polymer, thus providing evidence that the diffusion coefficient is dependent on the gas-polymer-matrix composition.

Experimental results show that sorbed gases interact with polymeric chains, inducing changes in the structural and dynamic properties of the polymer. These properties and the interchain forces controlling them determine many of the physical characteristics of the matrix, including the solubility and diffusion coefficients. These results are inconsistent with the assumptions and the physical interpretations implicit in the dual-mode sorption and transport model, and strongly suggest that the sorption and transport of gases in glassy polymers should be represented by concentration (composition) dependent solubility and diffusion coefficients.

0097–6156/83/0223–0089$06.50/0
© 1983 American Chemical Society

The transport of gas in polymers has been studied for over 150 years (1). Many of the concepts developed in 1866 by Graham (2) are still accepted today. Graham postulated that the mechanism of the permeation process involves the solution of the gas in the upstream surface of the membrane, diffusion through the membrane followed by evaporation from the downstream membrane surface. This is the basis for the "solution-diffusion" model which is used even today in analyzing gas transport phenomena in polymeric membranes.

In 1879, von Wroblewski (3) showed that sorption and transport of gases in polymers followed Henry's and Fick's laws, respectively,

$$C = \sigma o \ p \qquad (1)$$

$$J = -Do \ (dC/dx) \qquad (2)$$

where C is the equilibrium gas concentration in the polymer, J is the permeation flux, σo and Do are the solubility and diffusion coefficients, respectively, and p is the gas pressure. By assuming that Do is constant, von Wroblewski showed that the steady state flux is given by,

$$J = \sigma o \ Do \ (p/\ell) = Po \ (p/\ell) \qquad (3)$$

where p is the pressure difference across the membrane (for simplicity, the downstream pressure has been assumed to be zero), and ℓ is the thickness of the membrane. The permeability coefficient at steady state, Po, is defined by eq. (3) as,

$$Po = \sigma o \ Do \qquad (4)$$

Daynes (4) (1920) showed that a transient permeation experiment, under the boundary conditions $C(x;t) = 0$ for $t<0$, $C(0;t) = C$ and $C(\ell;t) = 0$ for $t>0$ gives two characteristic parameters, the permeability, Po, and the time-lag, Θo. Daynes showed that the solution of Fick's law under these boundary conditions yields,

$$Do = \ell^2/6\Theta o \qquad (5)$$

Thus, Po, Do, and σo [by eq. (4)] could be measured in a single time-lag experiment.

Eqs. (1)-(5) are still the basic sorption and transport equations used today for "ideal" systems, penetrant-polymer systems in which both σo and Do are pressure and concentration independent. This "ideal" behavior is observed in sorption and transport of permanent and inert gases in polymers well above their Tg.

Barrer (6) (1937) showed that diffusion is an activated process and that the diffusion coefficient had an Arrhenius form,

$$Do = D\overset{*}{o} \exp(-Ed/RT) \tag{6}$$

where Ed is the activation energy of diffusion and $D\overset{*}{o}$ is the frequency or pre-exponential factor. A number of attempts have been made to explain the temperature dependence of the diffusion coefficient. The zone-theory of Barrer (7), the free volume theories (8, 9), and the molecular theories of diffusion (10, 11, 12), although differing in many aspects, have in common the connotation that the activation energy of diffusion is the energy needed for chain separation, through cooperative motions, of sufficient size to allow the penetrant to execute a diffusional jump.

Section IA summarizes the molecular model of diffusion of Pace and Datyner (12) which proposes that the diffusion of gases in a polymeric matrix is determined by the cooperative main-chain motions of the polymer. In Section IB we report carbon-13 nmr relaxation measurement which show that the diffusion of gases in poly(vinyl chloride) (PVC) - tricresyl phosphate (TCP) systems is controlled by the cooperative motions of the polymer chains. The correlation of the phenomenological diffusion coefficients with the cooperative main-chain motions of the polymer provides an experimental verification for the molecular diffusion model.
Section IIA summarizes the physical assumptions and the resulting mathematical descriptions of the "concentration-dependent" (5) and "dual-mode" (13) sorption and transport models which describe the behavior of "non-ideal" penetrant-polymer systems, systems which exhibit nonlinear, pressure-dependent sorption and transport. In Section IIB we elucidate the mechanism of the "non-ideal" diffusion in glassy polymers by correlating the phenomenological diffusion coefficient of CO_2 in PVC with the cooperative main-chain motions of the polymer in the presence of the penetrant. We report carbon-13 relaxation measurements which demonstrate that CO_2 alters the cooperative main-chain motions of PVC. These changes correlate with changes in the diffusion coefficient of CO_2 in the polymer, thus providing experimental evidence that the diffusion coefficient is concentration dependent.

I. DIFFUSION AND COOPERATIVE MAIN-CHAIN MOTIONS

A. Diffusion Theory

Recently Pace and Datyner (12) advanced a molecular theory of diffusion that correlates the diffusion of gases in a polymeric matrix with the cooperative motions of the polymer chains. The theory proposes that the diffusant molecule can move through the polymer matrix in two distinct ways: (a) sliding along the axis of interchain channels or bundles formed by adjacent polymer chains, or (b) jumping at right angles to the polymer chains whenever adjacent chains are sufficiently separated. The first process has a smaller activation energy than the second and therefore occurs much more rapidly. This is true because only limited main-chain motions are necessary to create interchain channels of molecular dimensions in disordered glasses. The penetrant diffuses freely within these interchain channels, which may be bounded by regions of dense local packing. The diffusant can proceed to the next region of facile diffusion only when two adjacent polymer chains undergo sufficient cooperative motions to cause an interchain separation greater than the molecular dimensions of the diffusant. Thus, process (b) is the rate determining step in diffusion and the activation energy of diffusion is the activation energy of the chain separation. In other words, the phenomenological diffusion process is the result of two dynamic events: the first, high-frequency motions ($>10^{10}$ Hz) of the diffusant within interchain channels; and the second, low-frequency, cooperative motions (10^4-10^8 Hz) of the polymer chains to allow the diffusant to enter another interchain channel. The former motions have been observed experimentally in diffusion experiments ($\underline{14}$, $\underline{15}$), the latter are reported here.

The activation energy for a symmetrical separation of polymer chain centers by a distance of z over a length x is given by ($\underline{12}$),

$$\Delta E = \int_{-\infty}^{\infty} \left[f(z) - f(\rho) + \frac{\beta}{2} \left[\frac{d^2z}{dx^2}\right]^2 \right] dx \qquad (7)$$

where $f(z)$ is the average interchain potential per unit length, β is the average effective single chain-bending modulus per unit length, and ρ is the equilibrium chain separation. Solution of eq. (7) for $z = d$, where d is the minimum chain separation which will allow transverse passage of a penetrant, gives the activation energy of diffusion,

$$Ed = \Delta E = 3.11 \; \Gamma^{3/4} \; \beta^{1/4} \; d^{5/4} \qquad (8)$$

where Γ and β are parameters which characterize the interchain cohesion and chain stiffness, respectively. The value of Γ can be estimated from polymer density and cohesive energy density, and β can be approximated from the polymer chain backbone geometry and bond rotation potentials. The theory makes the unique prediction that the activation energy of diffusion depends nearly linearly on the penetrant diameter. This prediction has been recently confirmed experimentally by Berens and Hopfenberg ($\underline{16}$).

From this molecular theory, we see that the diffusion coefficient depends on the frequency of cooperative main-chain motions of the polymer, ν, which cause chain separations equal to or greater than the penetrant diameter. Pace and Datyner were able to estimate ν by adopting an Arrhenius rate expression in which the pre-exponential factor, A, is a function of both ΔE and T. The diffusion coefficient is given by,

$$Do = \frac{1}{6} \bar{L}^2 \, \nu = \frac{1}{6} \bar{L}^2 \, A \, \exp(-\Delta E/RT) \qquad (9)$$

where $\bar{L}^2$ is the mean-square jump displacement. The characteristic frequency of the diffusion process establishes a time scale ($\tau = 1/\nu$) for jumps of the diffusant molecule in the polymeric matrix. Hence, $\bar{L}$ is the mean displacement in the average position of the gas molecule following an activated jump. (The average position of a penetrant molecule is determined by its random motion within the region of facile diffusion, process (a), and averaged over time τ.) The average jump distance, $\bar{L}$, is not predictable within the limits of the theory.

B. Diffusion Mechanism in Polymers - Experimental Evidence

1. Cooperative Main-Chain Motions in PVC-TCP

It is now well established that TCP acts as an antiplasticizer of PVC at low concentrations and as a plasticizer at high concentrations. The observed variations in tensile modulus, tensile strength, impact strength, ultimate elongation, and thermal expansion coefficient of PVC with additive concentration have been attributed to the antiplasticization-plasticization phenomenon (17, 18, 19). Antiplasticization of PVC also results in a decrease in gas permeability, and plasticization in an increase in gas permeability (17).
Based on changes of secondary-loss transitions of antiplasticized polymers, Robeson (20) speculated that additives altered the cooperative motions of the polymer chains, and these alterations were responsible for the observed changes in diffusivity. In particular, Robeson suggested that the addition of antiplasticizers to polymers restricted molecular flexibility of chains thereby restricting diffusion of penetrants. Kinjo (18) attributed the antiplasticization effect of several polar low-molecular weight additives in PVC to cohesion of the additive to the polymer chain by dipole-dipole interactions. He proposed that motions of the cohering parts of the PVC would be extremely hindered, reducing the mechanical β-dispersion and increasing tensile strength. It was suggested in the literature (21) that high concentrations of TCP in PVC decrease the interchain

interactions in the polymer, thus enhancing the main-chain motions, resulting in plasticization of the polymeric matrix.

We have been interested in the nature of cooperative motions in polymers for some time and have used carbon-13 nuclear magnetic resonance for examining main-chain motions in solids (22-27). Carbon-13 nmr with cross-polarization and magic-angle sample spinning is a high-resolution, high-sensitivity technique for solids (Figure 1). The cooperative main-chain motions in glassy polymers can be evaluated from carbon-13 rotating-frame relaxation rates, $[R_1\rho(C)]$. Details of the nmr experiment are reported elsewhere (28).

The average relaxation rate of the methylene- and methine-carbons in PVC decreases with the addition of TCP up to about 15 weight % (Table I). Based on standard relaxation rate theory (29), reduced relaxation rates are indicative of a shift in the average cooperative motions to lower frequencies. These results prove that antiplasticization is associated with reduced cooperative motions of the polymer chains. As the concentration of TCP in PVC is increased above 15 weight %, the average relaxation rate of the methylene- and methine-carbons in PVC increases (Table I), showing that plasticization leads to increased cooperative motions in the glass.

Our results show for the first time the effect of antiplasticization-plasticization on polymer properties at the molecular level. These results provide experimental evidence for the assumptions of Robeson (20) and Kinjo (18) that attributed the changes in polymer properties with antiplasticization-plasticization to changes in the interchain cohesion of the polymer.

2. Transport in PVC-TCP

Hydrogen and carbon monoxide permeability data for films of PVC and PVC containing TCP are given in Table I. The permeability coefficients, of both gases, decrease as TCP concentration is increased to about 15 weight %, and then increase sharply with additive concentration. Similar results have been reported for CO_2 and H_2O permeabilities in the PVC-TCP system (17).

Table I lists also the apparent diffusion coefficients, Da, calculated from the time lag, θ. The dependence of the apparent diffusion coefficients on additive concentration is similar to the dependence of the permeability coefficients on this concentration. We will show later that the diffusion coefficients in PVC are dependent on the gas concentration in the polymer, and we emphasize here that the values quoted in Table I are the apparent diffusion coefficients and are not the values of either the real or effective diffusion coefficients, D and $\bar{D}$, respectively. We recognize that Da actually underestimates both D and $\bar{D}$, but we can still use the apparent diffusion coefficients to show the

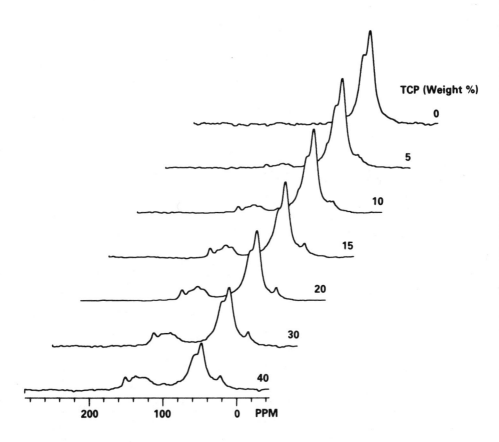

Figure 1. Cross-polarization and magic-angle spinning ^{13}C-NMR spectra of PVC-TCP systems. The 15.08 MHz spectra were obtained with 1 msec contact time and 1800 Hz spinning speed.

INDUSTRIAL GAS SEPARATIONS

TABLE I

GAS TRANSPORT AND C-13 NMR PARAMETERS FOR PVC-TCP

Additive (weight %) TCP	$P \times 10^{10}$ CC(STP)·cm/cm²·sec·cmHg[a]		$D_a \times 10^7$ cm²·sec^{-1}[a]		$\langle R_{1\rho}(C) \rangle \times sec^{-1}$[b]
	H_2	CO	H_2	CO	
0	2.38 ± .06	.0248 ± .0004	4.80 ± .20	.0229 ± .0012	235
5.0	1.91 ± .06	.0154 ± .0015	4.56 ± .37	.0210 ± .0006	170
10.2	1.77 ± .01	.0178 ± .0003	4.22 ± .33	.0157 ± .0001	140
15.0	1.69 ± .01	.0180 ± .0002	4.44 ± .23	.0251 ± .0002	120
20.1	2.16 ± .04	.0274 ± .0016	4.73 ± .58	.0293 ± .0004	155
30.8	2.86 ± .02	.1050 ± .0010	15.6 ± 2.20	.0535 ± .0023	250
40.0	3.61 ± .01	.3740 ± .0030	22.5 ± 1.40	.2880 ± .0100	430

a. At 270 cmHg and 27°C.
b. Measured at 34 kHz and 26°C.   Average relaxation rate for methylene and methine carbons.

relative changes in these diffusion coefficients as the changes in D and $\bar{D}$ are proportional to the changes in Da. Further, both Barrer (30) and Florianczyk (31) have shown that the solubility of gases in PVC is only weakly dependent on TCP concentration. Thus, the changes in the permeabilities must be due mainly to changes in the effective diffusion coefficients [see eq. (12)].

3. Diffusion and Cooperative Main-Chain Motions in PVC-TCP

The molecular theory of Pace and Datyner (12) predicts that the frequency of polymer motions important to diffusion of CO and H_2 in PVC is in the range of 10^5-10^8 Hz. We can expect $R_1\rho(C)$ measurements performed between 10^4 to 10^5 Hz to be sensitive to alterations in ν by additives. The dependence of $<R_1\rho(C)>$ on the rotating-frame Larmor frequency observed for the PVC-TCP system (28) means that a general change in main-chain motions (restriction or enhancement) will result in a change (decrease or increase, respectively) in $<R_1\rho(C)>$ measured at 34 kHz (29).

When the concentration of additive is low we observe that the Da of both gases decrease and parallel the decrease in $<R_1\rho(C)>$ (Fig. 2). The addition of low levels of TCP increases the average interchain potential in PVC and results in a decrease in ν, and in lower diffusion coefficients. At high concentrations of TCP both $<R_1\rho(C)>$ and Da increase (Fig. 2). The dilution of the chains by the low molecular weight additive decreases the interchain potential, thereby increasing the frequency of main-chain motions, and increasing the diffusion coefficients.

In conclusion, the average rotating-frame relaxation rate of the methylene- and methine-carbons correlate with the apparent diffusion coefficients for H_2 and CO in PVC when the main-chain molecular motions of the polymer are altered by an additive. (Fig. 2). These results provide experimental evidence that main-chain cooperative motions control the diffusion of gases through polymers. In Section IIB we will show that perturbation of polymeric cooperative motions is not restricted to classical plasticizing additives.

II. SORPTION AND TRANSPORT IN GLASSY POLYMERS

Nonlinear, pressure-dependent solubility and permeability in polymers have been observed for over 40 years. Meyer, Gee and their co-workers (5) reported pressure-dependent solubility and diffusion coefficients in rubber-vapor systems. Crank, Park, Long, Barrer, and their co-workers (5) observed pressure-dependent sorption and transport in glassy polymer-vapor systems. Sorption and transport measurements of gases in glassy polymers show that these penetrant-polymer systems do not obey the "ideal" sorption and transport eqs. (1)-(5). The observable variables,

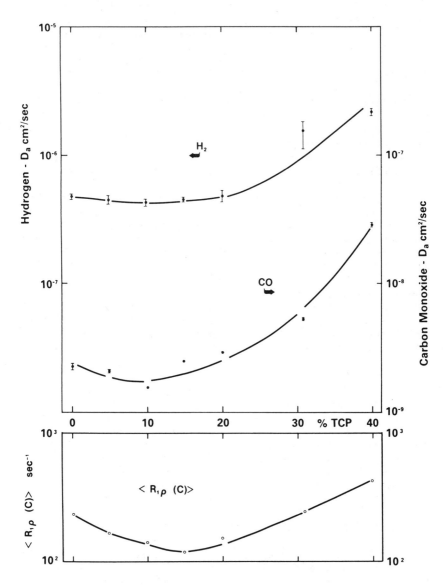

Figure 2. The dependence of the main-chain rotating-frame relaxation rate, and apparent diffusion coefficients of H_2 and CO, on the concentration of TCP in PVC. The relaxation rates were measured at 34 kHz and 26 °C. The diffusion coefficients were measured at 270 cm-Hg and 27 °C.

solubility coefficient, permeability, and diffusion time-lag are pressure dependent (13).

A number of attempts have been made to explain the nonlinear, pressure-dependent sorption and transport in polymers. These explanations may be classified as "concentration-dependent" (5) and "dual-mode"(13) sorption and transport models. These models differ in their physical assumptions and in their mathematical descriptions of the sorption and transport in penetrant-polymer systems.

A. Sorption and Transport Models

1. Concentration-Dependent Sorption and Transport Model

Pressure-dependent sorption and transport properties in polymers can be attributed to the presence of the penetrant in the polymer. Crank (32) suggested in 1953 that the "non-ideal" behavior of penetrant-polymer systems could arise from structural and dynamic changes of the polymer in response to the penetrant. As the properties of the polymer are dependent on the nature and concentration of the penetrant, the solubility and diffusion coefficient are also concentration-dependent. The concentration-dependent sorption and transport model suggests that "non-ideal" penetrant-polymer systems still obey Henry's and Fick's laws, and differ from the "ideal" systems only by the fact that σ and D are concentration dependent,

$$C = \sigma \, p = \sigma o \, [1 + g(C)] \, p \qquad (10)$$

$$J = -D \, (dC/dx) = -Do \, [1 + f(C)] \, (dC/dx) \qquad (11)$$

where $g(C)$ and $f(C)$ are functions describing the concentration dependence of σ and D, respectively. Solution of eq. (11) under the boundary conditions of the transient permeation experiment yields,

$$P = \bar{D} \, \sigma = Do \, [1 + f'(C)] \, \sigma o \, [1 + g(C)]$$

$$= Po \, [1 + f'(C)] \, [1 + g(C)] \qquad (12)$$

$$\theta = (\ell^2/6Do) \, [1 + F(C)] = \theta o \, [1 + F(C)] \qquad (13)$$

where $\bar{D}$, the effective diffusion coefficient, is defined by $\bar{D} = (1/C) \int DdC$, and the functions $f'(C)$ and $F(C)$ describe the concentration dependence of $\bar{D}$ and θ, respectively.

Crank, Park, Long, Barrer and their co-workers (5) have shown that $\bar{D}$ and D can be represented as exponential functions of penetrant concentration. Aitken and Barrer (33) used successfully a linear expression to describe the concentration

dependence of D. Frisch ($\underline{34}$) developed explicit expressions for
θ for cases where D is concentration-dependent.

2. Dual-Mode Sorption and Transport Model

Barrer, Barrie, and Slater ($\underline{35}$) suggested in 1958 that
nonlinear sorption isotherms of gases and vapors in glassy
polymers could be explained by invoking two distinct mechanisms
of sorption. They described this dual-mode sorption as arising
from ordinary dissolution (Henry's law mode) plus absorption in
pre-existing "holes" (Langmuir's mode). Michaels, Vieth, and
Barrie ($\underline{36}$) suggested in 1963 that only the dissolved gas
(Henry's population) was involved in the transport process.
Vieth and Sladek ($\underline{37}$) extended this idea and developed the
total-immobilization model of gas transport. The transport
model, assuming complete immobilization of the absorbed gas
(Langmuir's population), was modified by Petropoulos ($\underline{38}$) and
later by Paul and Koros ($\underline{39}$). They suggested that the Langmuir
population was not immobilized but had a finite mobility. The
dual-mode sorption and transport model describes the sorption of
a gas in a glassy polymer by a combination of Henry's and
Langmuir's isotherms, eq. (14), and the gas transport by Fick's
law, eq. (15),

$$C = C_D + C_H = k_D p + \frac{C_H' bp}{1 + bp} \qquad (14)$$

$$J = -D_D \frac{dC_D}{dx} - D_H \frac{dC_H}{dx} \qquad (15)$$

where C_D and C_H are the concentrations of Henry's and Langmuir's
populations, respectively, D_D and D_H are the respective diffusion
coefficients of the two populations, C_H' and b are the capacity
and affinity parameters of the Langmuir sorption, respectively,
and k_D is Henry's law solubility coefficient. Solution of eq.
(15) for the boundary conditions of the transient permeation
experiment yields ($\underline{39}$),

$$P = k_D D_D \left(1 + \frac{KF}{1 + bp}\right) \qquad (16)$$

$$\theta = \frac{\ell^2}{6D_D} [1 + f(K, F, bp)] \qquad (17)$$

where $K = C_H' b / k_D$, $F = D_H / D_D$, and the function symbolized is too
cumbersome to reproduce here.

In summary, the dual-mode sorption and transport model assumes (13):
(a) Two modes of sorption - normal Henry's mode dissolution, and Langmuir's mode absorption into unrelaxed-volume frozen in the glassy state.
(b) The two gas populations are in a dynamic equilibrium with each other.
(c) The two gas populations have constant but different diffusion coefficients. Langmuir's population generally has considerably less diffusional mobility than Henry's population.
(d) Gases, particularly permanent gases, do not interact with the polymer matrix, thus Henry's law solubility coefficient, k_D, and the diffusion coefficients of both Henry's population, D_D, and Langmuir's population, D_H, are pressure and concentration-independent.

In the dual-mode sorption and transport model the pressure-dependence of σ (= C/p), P and Θ in gas-glassy polymer systems arises from the pressure-dependent distribution of the sorbed gas molecules between Langmuir sites and Henry's law dissolution. Although k_D, D_D and D_H are assumed to be constant, the average or effective solubility and diffusion coefficients of the entire ensemble of gas molecules change with pressure as the ratio of Henry's to Langmuir's population, C_D/C_H, changes continuously with pressure [eq. (14)].

B. Diffusion Mechanism in Glassy Polymers - Experimental Evidence

1. Cooperative Main-Chain Motions in PVC-CO$_2$

In Section IB we showed that carbon-13 rotating-frame relaxation measurements can be used to measure cooperative main-chain motions in polymers (28). We report here the effect of CO$_2$ on the main-chain motions of PVC.

Carbon-13 rotating-frame relaxation rates $\langle R_1\rho(C)\rangle$, were determined on static samples using cross-polarization techniques (40). Without magic-angle sample spinning, the methylene- and methine-carbon resonances are unresolved, so the relaxation rates reported here are the average relaxation rates for the combination line. The $\langle R_1\rho(C)\rangle$, at 37 kHz, for PVC in vacuo and in the presence of CO$_2$ are given in Table II. Two major effects are observed; the presence of relatively small amounts of CO$_2$ increases the $\langle R_1\rho(C)\rangle$ of PVC, and exposure of PVC to CO$_2$ causes a long-term increase in $\langle R_1\rho(C)\rangle$. The first phenomena will be discussed in terms of its effect on transport properties, and the second in terms of history-dependent properties.

The $\langle R_1\rho(C)\rangle$ for "conditioned" PVC in vacuo is 154 sec^{-1}. The presence of CO$_2$ leads to noticeable increases in $\langle R_1\rho(C)\rangle$. At 100 torr the $\langle R_1\rho(C)\rangle$ of PVC is increased to 158 sec^{-1}.

TABLE II

Relaxation Rates in PVC-CO_2

Sequence of Experiment	Sample Description	Relaxation Rate[a] $<R_{1\rho}(C)>$ sec^{-1}
1	in vacuo	138 ± 3
3	degassing[b] after first exposure to 800 torr CO_2-in vacuo	151 ± 4
7	degassing after repeated exposure to CO_2-in vacuo	154 ± 3
5	100 torr CO_2[c]	158 ± 5
6	200 torr CO_2	163 ± 8
4	400 torr CO_2	166 ± 3
2	800 torr CO_2	183 ± 4

a) Least-squares fit of relaxation data collected for times between 0.05 and 1 msec; error limits are one standard deviation; $H_1(C)$ = 37 kHz. Average for methylene and methine carbons.

b) All samples were degassed for 24 hours at 10^{-3} torr.

c) Measured after 12 hours equilibration at stated pressure. The concentration of CO_2 ranges from about 1 CO_2 molecule per 1200 repeat units at 100 mmHg to about 1 CO_2 per 200 repeat units at 800 mmHg (19).

Higher pressures of CO_2 cause correspondingly larger increases in $<R_1\rho(C)>$. At 800 torr an increase of 19% to 183 sec^{-1} is observed. Based on standard relaxation rate theory (29), increased relaxation rates are indicative of a shift in the average cooperative main-chain motions to higher frequencies. Conversely, this means that even small amounts of CO_2 increase the cooperative motions of the polymer chains.

Gravimetric measurements by Berens (41) show that CO_2 sorption by PVC ranges from about 0.5 mg CO_2/gram of PVC at 100 torr to about 3.5 mg CO_2/gram of PVC at 800 torr of CO_2. The reason that such small amounts of CO_2 are so effective in altering the cooperative motions of PVC results from the fact that gases have very high molecular mobilities in glassy polymers (correlation time between 10^{-10}-10^{-12} sec) and can sample extended areas of the polymer on the time scale of the cooperative motions of the polymer (10^{-5}-10^{-8} sec).

The experimental data in Table II shows that $<R_1\rho(C)>$ of PVC becomes longer after successive exposures of the sample to CO_2. This indicates that exposure to carbon dioxide causes a change in the molecular packing of PVC which allows more main-chain motions. There is now abundant evidence that exposure of polymers to gases results in changes which persist long after the gases are removed. These changes are reflected in altered physical properties of the polymer (42-46). The "conditioning" of a polymer by exposure to a gas can be thought of in the same terms as annealing. In annealing, the increased thermal energy allows the polymer sufficient segmental mobility to eliminate energetically unfavorable conformations which were frozen into the solid on rapid cooling from the melt (47). Exposure to a gas can have the opposite effect. Since the polymer must swell to accommodate the sorbed gas molecules, the polymer becomes stressed and ultimately reaches a new equilibrium condition determined by the presence of the penetrant. The concentration and the nature of the penetrant determine the rate of relaxation just as temperature determines the rate of relaxation in the annealing experiment.

Since the penetrant enhances cooperative motions of the polymer, the rate of structural change in the polymer, measured by various relaxation experiments, will be greater during sorption, or in the presence of the penetrant, than it will be during desorption, or in the absence of the penetrant (32). Because of this, "conditioning" of a polymer results in long-term or "permanent" changes in the polymer. The results of sorption and transport experiments therefore depend on the static and dynamic state of the polymer, which are determined by the thermal history of the sample and by the pressure and duration of previous exposures to sorbents. Wonders and Paul (42) have described these phenomenological effects in detail for the CO_2-polycarbonate system.

2. Diffusion and Cooperative Main-Chain Motions in PVC-CO$_2$

In Section IB we presented experimental evidence that diffusion coefficients correlate with PVC main-chain polymer motions. This relationship has also been justified theoretically (12). In the previous section we demonstrated that the presence of CO$_2$ effects the cooperative main-chain motions of the polymer. The increase in <R$_1\rho$(C)> with increasing gas concentration means that the real diffusion coefficient [D in eq. (11)] must also increase with concentration. The nmr results reflect the real diffusion coefficients, since the gas concentration is uniform throughout the polymer sample under the static gas pressures and equilibrium conditions of the nmr measurements. Unfortunately, the real diffusion coefficient, the diffusion coefficient in the absence of a concentration gradient, cannot be determined from classical sorption and transport data without the aid of a transport model. Without prejustice to any particular model, we can only use the relative change in the real diffusion coefficient to indicate the relative change in the apparent diffusion coefficient.

Apparent diffusion coefficients of CO$_2$ in PVC have been reported by Toi (48). Semilog plots of both the apparent diffusion coefficient, Da, and the relaxation rate, <R$_1\rho$(C)>, as a function of CO$_2$ pressure in the range of 0-800 torr are shown in Figure 3. Of course, there is no reason to expect the magnitude of the change in the rotating-frame relaxation rate to be equal to the magnitude of the change in D or Da. Both the relaxation rate and the diffusion coefficient depend on the frequency of the main-chain cooperative motions but in different ways (28, 29, 49). Nevertheless, the similarity in the dependence of both the relaxation rate and the apparent diffusion coefficient on CO$_2$ pressure supports the observation that main-chain molecular motions play a major role in determining the diffusion coefficient of a gas through a polymer matrix (28).

The experimental evidence presented here and in the literature (15) show that the real diffusion coefficient depends on concentration. These results are incompatible with the notion of concentration-independent diffusion coefficients for the dissolved and Langmuir sorbed molecules [D$_D$ and D$_H$ in equation (15)] as proposed by the dual-mode sorption and transport model (13).

C. Concentration-Dependent Sorption and Transport
    in Glassy Polymers

The basic difference between "concentration-dependent" and "dual-mode" models is in their assumption about penetrant-polymer interactions. Concentration-dependent sorption and transport models are based on the assumption that the concentration-dependence of the solubility and diffusion coefficients arises

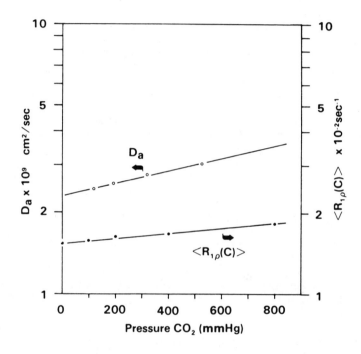

Figure 3.   The dependence of relaxation rate and apparent diffusion coefficient on CO_2 pressure in PVC.   The relaxation rates were measured at 37 kHz and 26 °C.   The diffusion coefficients, measured at 40 °C, are from Ref. 48.

from penetrant-polymer interactions. On the other hand, the dual-mode sorption and transport model assumes that there are no gas-polymer interactions, thus assuming concentration-independent solubility and diffusion coefficients - k_D, D_D and D_H, respectively.

Experimental results presented in this work and in the literature are inconsistent with the assumptions and the physical interpretations implicit in the dual-mode sorption and transport model, and strongly suggest that the sorption and transport in gas-glassy polymer systems should be presented by a concentration-dependent model:

(a) Evidence for gas-glassy polymer interactions is abundant. The results presented in Table II show that even small amounts of gas affect the cooperative main-chain molecular motions of glassy polymers. Evidence that the presence of gases in polymer cause structural and dynamic changes can be seen in the depression of the Tg (42, 43, 44), and in the increased viscoelatic relaxation rates (43, 44) of polymer-gas systems. Further, "conditioning" of polymers, the gas-induced structural changes in polymers by exposure to penetrant, was shown by Wonders and Paul (42), as well as by our work (40) (Table II).

(b) The effect of gas on the cooperative main-chain motions of glassy polymers (Table II) shows that the molecular level diffusion process is concentration dependent, as must be the phenomenological diffusion coefficient.

(c) There is only one population of sorbed gas in a glassy polymer at any given pressure. All spectroscopic analyses (15, 22), including the work of Assink (14), are consistent with all of the sorbed gas molecules being in a single state. Nuclear magnetic spin-spin and spin-lattice relaxation measurements of sorbed gases consistently show single exponential decays indicating that all of the gas molecules are relaxed by the same mechanism. These studies also show that the relaxation times increase with increasing equilibrium gas pressure. There are at least two possible explanations for these results: 1) there is a single population of gas molecules which interact with the polymer matrix; or 2) there are two populations with different mobilities in rapid exchange and whose relative populations vary with pressure. The results presented in Section IIB are consistent only with the first explanation (40).

The work of Assink (14) has been frequently cited as proof for the existence of two distinct sorption modes in glassy polymers (50). This is not the case. Measuring the spin-spin relaxation time (T_2) of ammonia in polystyrene, Assink showed that the ammonia is relaxed by a single exponential process and that T_2 increases with increasing concentration of sorbed ammonia. Assink eliminated the

obvious   interpretation   of   gas-polymer   interactions   by
assuming that at the experimental pressures of up to 6 atm.
"the concentration of dissolved gas was not sufficient to
appreciably   plasticize   the   polymer".   He   proceeded   to
analyze his experimental results in terms of the dual-mode
sorption   theory   by   attributing   different   mobilities   and
therefore different T_2's to Henry's and Langmuir's populations. Assink plotted T_2 as a function of the mole fraction
of the gas in the Langmuir sites, as determined from the
dual-mode model, so the slope of the line should give $1/T_2$
for the Langmuir fraction of the ammonia and the intercept
would be the $1/T_2$ for the dissolved or Henry's fraction of
the gas. A least-squares analysis of the data gave a slope
of 235 sec^{-1} and an intercept of -5.2 sec^{-1}. A negative
relaxation   rate   is   physically   impossible.   Rather   than
abandoning   the   attempt   to   represent   the   data   by   dual-
sorption-mobility model, Assink arbitrarily assigned a valve
of 12 sec^{-1} for $1/T_2$ of the dissolved molecules making the
mobility   of   the   Langmuir   molecules   1/20th   that   of   the
dissolved molecules.
The results of Zupančič, et.al. (15), are also relevent
here.   Zupančič   investigated   diffusion of butane in linear
polyethylene   using   pulsed   magnetic   field   gradient
experiments   to   measure   directly   the   real   diffusion
coefficient at 23°C. They showed that the real diffusion
coefficient increased with equilibrium butane pressure as
did the spin-spin relaxation time. As the polyethylene was
above its Tg, these results cannot be attributed to dual
mode   behavior   (13).   A   single   population   of   butane
interacting with the polyethylene accounts for both the
change   in   the   diffusion   coefficient   and   the   change   in
spin-spin relaxation.
The only exception to a single population of sorbent in a
glassy polymer was observed in the water-cellulose acetate
system (51, 52). In this system two resonance frequencies
and two relaxation rates for water were observed. However,
the two dynamic states of the water in this system are due
to specific hydrogen bonding interactions rather than sorption in Langmuir type holes.

(d)   There is no model-independent or physical evidence for the
existence of Langmuir or other sorption sites in glassy
polymers which could account for multiple-site sorption.
The interpretation of free volume as molecular-size voids
responsible   for   Langmuir   sorption   are   only   partially
successful.   While   free   volume   has   been   equated   with   the
Langmuir capacity, C'_H, in CO_2-glassy polymer systems (13,
46, 50, 53) the correlation fails to hold for other gases
(54, 55). Koros et. al. (50) explain the equivalence of
the   "unrelaxed   volume"   to   the   Langmuir   capacity   by

suggesting that the total "unrelaxed volume" is uniformly distributed as molecular scale gaps and that these gaps are Langmuir sorption sites for single CO_2 molecules. This physical model implies that the average Langmuir site has the same molecular volume as the effective molecular volume of CO_2 molecules when sorbed on zeolites or in liquids (80 $Å^3$) (50, 54). Methane is slightly larger than CO_2 so it is reasonable that not all sites accessible to CO_2 would be accessible to CH_4. Indeed, the Langmuir saturation capacities of polycarbonate (54) and polysulfone (55) for CH_4 are about 1/2 the C_H' of CO_2 capacities. However, argon and nitrogen are both slightly smaller than CO_2, so it should be possible to sorb one of these molecules into each Langmuir site under saturation conditions, but the C_H''s for Ar and N_2 in polycarbonate (54) and polysulfone (55) are again less than 1/2 the C_H' for CO_2 in the same polymers. To accept the physical rational of the dual-mode model would imply that CO_2 is unique among all penetrants in probing the "unrelaxed volume" of polymers. The discrepancy in the values of C_H' for different gases cannot be attributed to the condensibility of the penetrants, since condensibility determines the relative efficiency with which the penetrant can utilize the available volume and so would effect only the value of the Langmuir affinity parameter, b (13).

(e) Dilatometric measurements (56) of the swelling of "conditioned" polycarbonate by CO_2 show that the relative swelling of the polymer drops dramatically as the CO_2 pressure is increased. These results cannot be reconciled with the dual-mode sorption model, which assumes preferential sorption into pre-existing sorption sites at low pressures (sorption that should have little effect on the polymer dimensions) and mainly Henry's type dissolution at high pressures (dissolution that should result in the swelling of the polymer).

The dual-mode sorption and transport model provides an adequate mathematical description of the phenomenological sorption and transport in glassy polymers (13). However, the results presented here invalidate the physical assumptions implicit to the dual-mode model regarding no penetrant-polymer interactions and the existence of two distinct populations of sorbed gas molecules. In the initial formulation of the dual-sorption-mobility model, the authors cautioned against attaching physical significance to the mathematical parameters (39, 57). Since that time no physical evidence has appeared to support the existence of Langmuir sorption sites or concentration-independent diffusion and solubility coefficients in glassy polymers. Nevertheless, the mathematical parameters describing these two states have taken on a physical significance out of proportion with reality.

In the following chapter we present the matrix model of gas sorption and diffusion in glassy polymers which is based on the observation that gas molecules interact with the polymer, thereby altering the solubility and diffusion coefficients of the polymer matrix.

Literature Cited

1. Barr, G. "Dictionary of Applied Physics"; R. Glazerbrook, Ed.; MacMillan: London, 1923; Vol. 5.
2. Graham, T. Phil. Mag. 1866, 32, 401.
3. Von Wroblewski, S. Wiedemanns Ann. Physik. 1979, 8, 29.
4. Daynes, H. A. Proc. Roy. Soc. (London) A. 1920, 97, 286.
5. Barrer, R. M. J. Phys. Chem. 1957, 61, 178, and references therein.
6. Barrer, R. M. Nature 1937, 140, 106.
7. Barrer, R. M. Trans. Faraday Soc. 1939, 35, 644.
8. Bueche, F. J. Chem. Phys. 1953, 21, 1850.
9. Cohen M. H.; Turnbull, D. J. Chem. Phys. 1959, 31, 1164.
10. Brandt, W. W. J. Phys. Chem. 1959, 63, 1080.
11. DiBenedetto, A. T. J. Polym. Sci. A 1963, 1, 3477.
12. Pace, R. J.; Datyner, A. J. Polym. Sci., Polym. Phys. Ed. 1979, 17, 437.
13. Paul, D. R. Ber. Bunsenges Phys. Chem. 1979, 83, 294.
14. Assink, R. A. J. Polym. Sci., Polym. Phys. Ed. 1975, 13, 1665.
15. Zupančič, I.; Lahajnar, G.; Blinc, R.; Reneker, D. H.; Peterlin, A. J. Polym. Sci., Polym. Phys. Ed. 1978, 16, 1399.
16. Berens, A. R.; Hopfenberg, H. B. J. Membrane Sci. 1982, 10, 283.
17. Kinjo, N. Japan Plastics 1973, 7(4), 6.
18. Kinjo, N.; Nakagawa, T. Polymer Journal 1973, 4, 143.
19. Jacobson, U. British Plastics 1959, 32, 152.
20. Robeson, L. M. Polym. Eng. Sci. 1969, 9, 277.
21. Gardon, J. L. J. Colloid Interface Sci. 1977, 59, 582.
22. Schaefer, J.; Sefcik, M. D.; Stejskal, E. O.; McKay, R. A. Macromolecules, in press.
23. Schaefer, J.; Stejskal, E. O.; Buchdahl, R. Macromolecules 1977, 10, 384.
24. Steger, T. R.; Schaefer, J.; Stejskal, E. O.; McKay, R. M. Macromolecules 1980, 13, 1127.
25. Schaefer, J.; Stejskal, E. O.; Steger, T. R.; Sefcik M. D.; McKay, R. M. Macromolecules 1980, 13, 1121.
26. Sefcik, M. D.; Schaefer, J.; Stejskal, E. O.; McKay, R. A. Macromolecules 1980, 13, 1132.
27. Stejskal, E. O.; Schaefer, J.; Steger, T. R. Symp. Faraday Soc. 1979, 13, 56.
28. Sefcik, M. D.; Schaefer, J.; May, F. L.; Raucher, D.; Dub. S. J. Polym. Phys., Polym. Phys. Ed. in press.

29. Bloombergen, N.; Purcell, E. M.; Pound, R. V. Phys. Rev.
     1948, 74, 679.
30. Barrer, R. M.; Mallinder, R.; Wong, P. S-L. Polymer 1967, 8,
     321.
31   Florianczyk, T. Polimery 1978, 23, 431.
32. Crank, J. J. Polym. Sci. 1953, 11, 151.
33. Aitken, A.; Barrer, R. M. Trans. Faraday Soc. 1955, 51, 116.
34. Frisch, H. L. J. Phys. Chem. 1957, 61, 93.
35. Barrer, R. M.; Barrie, J. A.; Slater, J. J. Polym. Sci.
     1958, 27, 177.
36. Michaels, A. S.; Vieth, W. R.; Barrie, J. A. J. Appl. Phys.
     1963, 34, 13.
37. Vieth, W. R.; Sladek, K. J. J. Coll. Sci. 1965, 20, 1014.
38. Petropoulos, J. H. J. Polym. Sci. A 1970, 2, 1797.
39. Paul, D. R.; Koros, W. J. J. Polym. Sci., Polym. Phys. Ed.
     1976, 14, 675.
40. Sefcik, M. D.; Schaefer, J. J. Polym. Phys., Polym. Phys.
     Ed. in press.
41. Berens, A. R. Polym. Eng. Sci. 1980, 20, 95.
42. Wonders, A. G.; Paul, D. R. J. Membrane Sci. 1979, 5, 63.
43. Wang, W. C.; Sachse, W.; Kramer, E. J. Bull. Am. Phys. Soc.
     1981, 26, 367.
44. Wang, W. C.; Sachse, W; Kramer, E. J. J. Polym. Sci., Polym.
     Phys. Ed. 1982, 20, 1371.
45. Robinson, C. Trans. Faraday Soc. 1946, 42B, 12.
46. Chan, A. H.; Paul, D. R. Polym. Eng. Sci. 1980, 20, 87.
47   Dunn, C. M. R.; Turner, S. Polymer 1974, 15, 451.
48. Toi, K. Polym. Engr. Sci. 1980, 20, 30.
49. Pace, R. J.; Datyner, A. J. Polym. Sci., Polym. Phys. Ed.
     1980, 18, 1169.
50. Chern, R. T.; Koros, W. J.; Sanders, E. S.; Chen, S. H.;
     Hopfenberg, H. B. Chapter in this book.
51. Belfort, G.; Scherfig, J.; Seevers, D. O. J. Coll.
     Interface Sci. 1974, 47, 106.
52. Almagor, E.; Belfort, G. J. Coll. Interface Sci. 1978, 66,
     146.
53. Koros, W. J.; Paul, D. R. J. Polym. Sci., Polym. Phys. Ed.
     1978, 16, 1947.
54. Koros, W. J.; Chan, A. H.; Paul, D. R. J. Membrane Sci.
     1977, 2, 165.
55. Erb, A. J.; Paul, D. R. J. Membrane Sci. 1981, 8, 11.
56. Sefcik, M. D.; Andrady, A. L.; Schaefer, J.; Desa, E. Polym.
     Prepr., Am. Chem. Soc., Div. Polym. Chem. in press.
57. Koros, W. J.; Paul, D. R.; Rocha, A. A. J. Polym. Sci.,
     Polym Phys. Ed. 1976, 14, 687.

RECEIVED January 12, 1983

Sorption and Transport in Glassy Polymers

Gas–Polymer–Matrix Model

DANIEL RAUCHER and MICHAEL D. SEFCIK

Monsanto Company, St. Louis, MO 63167

The gas-polymer-matrix model for sorption and transport of gases in polymers is consistent with the physical evidence that; 1) there is only one population of sorbed gas molecules in polymers at any pressure, 2) the physical properties of polymers are perturbed by the presence of sorbed gas, and 3) the perturbation of the polymer matrix arises from gas-polymer interactions. Rather than treating the gas and polymer separately, as in previous theories, the present model treats sorption and transport as occurring through a gas-polymer matrix whose properties change with composition. Simple expressions for sorption, diffusion, permeation and time lag are developed and used to analyze carbon dioxide sorption and transport in polycarbonate.

Nonlinear, pressure-dependent sorption and transport of gases and vapors in glassy polymers have been observed frequently. The effect of pressure on the observable variables, solubility coefficient, permeability coefficient and diffusion timelag, is well documented ($\underline{1}$, $\underline{2}$). Previous attempts to explain the pressure-dependent sorption and transport properties in glassy polymers can be classified as "concentration-dependent" and "dual-mode" models. While the former deal mainly with vapor-polymer systems ($\underline{1}$) the latter are unique for gas-glassy polymer systems ($\underline{2}$).

The concentration-dependent models attribute the observed pressure dependence of the solubility and diffusion coefficients to the fact that the presence of sorbed gas in a polymer affects the structural and dynamic properties of the polymer, thus affecting the sorption and transport characteristics of the system ($\underline{3}$). On the other hand, in the dual-mode model, the pressure-dependent sorption and transport properties arise from a

0097–6156/83/0223–0111$06.00/0

unique mechanism which assumes two distinct modes of sorption (2, 4). The dual-mode model assumes that the diffusant gas does not interact with the polymer matrix and does not induce changes in the polymer characteristics, thus the diffusion coefficients of both populations are assumed to be independent of gas pressure or concentration (4, 5, 6).

In the preceding chapter (7) we elucidated the mechanism of gas diffusion in glassy polymers by presenting evidence that the diffusion of gases through polymer matrices is controlled by the cooperative main-chain motions of the polymer (8), and that these motions are altered by the presence of penetrants (9). These results mean that sorbed gases alter the structure and dynamics of the polymer matrix, thus altering the sorption and transport characteristics of the system. We proposed that gas-polymer interactions reduce the interchain potential of the polymer, thus decreasing the activation energy of polymer chain separation, and therefore the activation energy of diffusion. The decrease in the activation energies results in higher frequencies of cooperative main-chain motions and in higher diffusion coefficients. The heat of sorption, and ultimately the solubility of gases, depend on similar interactions (12). Since the gas-polymer interactions alter the interchain energies, solubility coefficients also vary as the structure and dynamics of the polymer matrix change (10).

The experimental results presented in the preceding chapter and in the literature are inconsistent with the assumptions and the physical interpretation implicit in the dual-mode model and strongly suggest that the sorption and transport in gas-glassy polymer systems should be represented by a concentration-dependent type model.

In Section I we introduce the gas-polymer-matrix model for gas sorption and transport in polymers (10, 11), which is based on the experimental evidence that even permanent gases interact with the polymeric chains, resulting in changes in the solubility and diffusion coefficients. Just as the dynamic properties of the matrix depend on gas-polymer-matrix composition, the matrix model predicts that the solubility and diffusion coefficients depend on gas concentration in the polymer. We present a mathematical description of the sorption and transport of gases in polymers (10, 11) that is based on the thermodynamic analysis of solubility (12), on the statistical mechanical model of diffusion (13), and on the theory of corresponding states (14). In Section II we use the matrix model to analyze the sorption, permeability and time-lag data for carbon dioxide in polycarbonate, and compare this analysis with the dual-mode model analysis (15). In Section III we comment on the physical implication of the gas-polymer-matrix model.

I. GAS-POLYMER-MATRIX MODEL

A model for sorption and transport of gases in polymers has to specifically account for the fact that the presence of sorbed gases in the polymer modifies the matrix ($\underline{7}$, $\underline{9}$). In developing the matrix model we are guided by the physical evidence relating to the mechanism of sorption and transport. The matrix model is consistent with the following observations and assumptions:

(a) There is only one population of sorbed gas molecules in a polymer at any given pressure. Spectroscopic analyses of gas molecules within polymer matrices are consistent with all of the gas molecules being in a single state ($\underline{16}$, $\underline{17}$, $\underline{18}$). Further, there is no physical evidence for the existence of Langmuir or other sorption sites in glassy polymers which could account for multiple sorption sites.

(b) The physical properties of a polymer matrix are perturbed by the presence of a sorbed gas. In addition to the observed changes in cooperative main-chain motions of the polymer in the presence of a sorbed gas ($\underline{7}$, $\underline{9}$), changes in viscoelastic relaxation ($\underline{19}$, $\underline{21}$, $\underline{22}$), and Tg ($\underline{15}$, $\underline{20}$, $\underline{21}$, $\underline{22}$) have also been reported.

(c) Based on the effect of sorbed gas on the main-chain motions of the polymer ($\underline{7}$, $\underline{9}$) we propose that the perturbation of the polymer matrix arises from interactions between the gas molecules and polymer chains.

The matrix model ascribes the apparent concentration dependence of polymer properties to gas-polymer interactions, which affect the interchain potential energy of the polymer. The thermodynamic analysis of solubility of gases in polymers (10) shows that the gas-polymer interactions affect the free energy of solution of gases in polymers, thus controlling the solubility coefficient. Pace and Datyner proposed in their statistical mechanical model of diffusion of gases in polymers ($\underline{13}$) that the interchain potential energy controls the activation energy of chain separation, the activation energy of diffusion, and the diffusion coefficients. Chow ($\underline{23}$) showed by classical and statistical thermodynamics that gas-polymer interactions are manifested in the depression of the glass-transition temperature of the polymer.

It is possible to gain some insight into the effect of sorbed gas on the interchain potential energy, by analogy to the effect of temperature on interchain cohesion forces. Raising the temperature of the polymer, like increasing the sorbed gas concentration in the polymer, reduces the interchain potential, resulting in higher diffusion coefficient and in lower solubility coefficient.

A. Matrix Model Expressions for Sorption and Transport

1. Solubility Coefficient

The solubility coefficients of gases in glassy polymers are
not constant, but decrease with increasing concentration of the
gas in the polymers (2). Calculations of the isosteric enthalpy
of sorption in several gas-polymer systems confirm that the gas-
polymer affinity is reduced with increasing sorbed gas concentrations (24, 25, 26). The change in the isosteric enthalpy of
sorption is a result of the changes in the polymer matrix induced
by the presence of the sorbed gas.
    We have shown previously (10) that based on Gee's thermo-
dynamic analysis of gas solubility in polymers (12) and the
theory of corresponding states (14), the solubility coefficient,
σ, can be expressed as,

$$\sigma = \sigma_o \exp(-\alpha^* \kappa) \tag{1}$$

where σ_o is the solubility coefficient in the zero-concentration
limit, α^* is a constant relating the excess free energy of mixing
to the depression of the glass-transition temperature of the
polymer by the gas and κ is defined as,

$$\kappa = [Tg(o) - Tg(C)]/T \tag{2}$$

where $Tg(o)$ and $Tg(C)$ are the glass-transition temperatures of
the pure polymer and the polymer-gas mixture, respectively. Chow
(23) derived an explicit expression for the depression of the
glass-transition temperature of polymers by diluents. To a first
approximation, $[Tg(o) - Tg(C)]$ may be regarded as proportional to
the gas concentration in the matrix (11). Thus, eq. (1) can be
rewritten as,

$$\sigma = \sigma_o \exp(-\alpha C) \tag{3}$$

where α is a constant relating $\alpha^* \kappa$ to the concentration, C. Eq.
(3) can be simplified by expanding the exponential as a power
series and neglecting the higher order-terms,

$$\sigma = \frac{\sigma_o}{1 + \alpha C} \tag{4}$$

2. Diffusion Coefficient

We have shown in the preceding chapter (7) that the presence
of gas increases the cooperative main-chain motions of glassy
polymers (9). The diffusion model of Pace and Datyner (13)

suggests that increased frequency of the cooperative motions results in an increased diffusion coefficient. Based on the diffusion model of Pace and Datyner (13) and the theory of corresponding states (14) the diffusion coefficient, D, can be represented by (10),

$$D = Do \ (1 + \beta^*\kappa) \ \exp(\beta^*\kappa) \tag{5}$$

where Do is the diffusion coefficient in the zero-concentration limit, and β^* is a constant relating the excess activation energy of chain separation to the depression of the glass-transition temperature of the polymer by the gas. As the diffusion coefficient changes with concentration (or κ), the diffusivity of the gas, in a steady-state transport experiment, changes continuously through the membrane. Experimentally determined diffusion coefficients reflect the continuous change of the diffusivity and are therefore called effective diffusion coefficients (27). The effective diffusion coefficient can be expressed as a function of κ by,

$$\bar{D} = (1/\kappa) \int Dd\kappa \tag{6}$$

Substituting eq. (5) into eq. (6) and integrating between the boundary conditions of a steady-state experiment yields,

$$\bar{D} = Do \ \exp(\beta^*\kappa) \tag{7}$$

As before, by assuming linearity between κ and C, eq. (7) can be rewritten as,

$$\bar{D} = Do \ \exp(\beta C) \tag{8}$$

where β is a constant relating $\beta^*\kappa$ to C. Eq. (8) can be simplified by expanding the exponential as a power series and neglecting the higher-order terms,

$$\bar{D} = Do \ (1 + \beta C) \tag{9}$$

B. Matrix Model Parameters of Sorption and Transport

1. Constants σo and Do - Sorption and Transport in the Zero Concentration Limit.

In the limit of zero-concentration, gas-glassy polymer systems behave "ideally." As the gas concentration in the membrane approaches zero the solubility coefficient becomes constant with the value σo [eqs. (1), (3) and (4)]. In the same limit, the diffusion coefficients are constant and equal to the diffusion coefficient Do, [eqs. (5), (7), (8) and (9)]. As typical of limiting values, σo and Do have no correspondence to reality, but merely describe the behavior of the gas-polymer system as if there were no induced changes in the polymer matrix.

It should be kept in mind that σ_o and D_o are really mean values, recognizing the fact that glassy polymers are hetero-geneous and contain a distribution of structural and dynamic states (28). As such, these parameters are dependent on the nature of the penetrant and of the polymer (13), as well as the distribution of properties in the polymer.

2. Parameter α and β - Concentration Dependence of Sorption and Transport

The matrix model ascribes the apparent concentration depen-dence of the solubility and diffusion coefficients to gas-polymer interactions. The parameters α and β are proportionality con-stants, indicating the effectiveness of gas-polymer interaction in inducing changes in the solubility and diffusion coefficients. The explicit expressions for σ and D describe them as functions of the excess free energy of solution and the change in the activation energy of chain separation, respectively (10). As our understanding of the effect of the gas-polymer matrix composition on these excess energies increases we may find it possible to explicitly determine these energy parameters and thus σ and D without having to resort to empirically determined proportion-ality constants.

II. ANALYSIS OF SORPTION AND TRANSPORT DATA BY THE MATRIX MODEL

We show elsewhere (10) that sorption and transport expres-sions derived from eqs. (1) and (5) represent experimental sorption and transport data well. Nevertheless, the calculations are cumbersome and require additional polymer parameters. These disadvantages might hinder the broad use of these expressions; thus, we will base our following discussions on the simplified expressions, eqs. (4) and (9), exclusively.

In the derivation of the simplified expressions for solubil-ity and diffusion coefficients, eqs. (4) and (9), C was assumed to be small. This fact does not limit the usefulness of these expressions for high concentrations. We show below that sorption and transport expressions, eqs. (11) and (14), respectively, derived from the simplified equations retain the proper func-tional form for describing experimental data without being needlessly cumbersome. Of course, the values of the parameters in eqs. (4) and (9) will differ from the corresponding parameters in eqs. (3) and (8), to compensate for the fact that the trun-cated power series used in eqs. (4) and (9) poorly represent the exponentials when $\alpha C > 1$ or $\beta C > 1$. Nevertheless, this does not hinder the use of the simplified equations for making correlation between gas-polymer systems.

A. Sorption of Gases in Glassy Polymers

The concentration of sorbed gas in the polymer is the product of the solubility coefficient and the gas pressure p,

$$C = \sigma \; p \tag{10}$$

When σ is given by eq. (4)

$$C = \frac{\sigma o \; p}{1 + \alpha C} \tag{11}$$

The expression for the solubility, eq. (11), can be easily rearranged by solution of the quadratic equation to give a simple dependence of C on p,

$$C = \frac{(1 + 4\alpha \; \sigma o \; p)^{\frac{1}{2}} - 1}{2\alpha} \tag{12}$$

The solid line in Fig. 1 represents the sorption isotherm of carbon dioxide in polycarbonate calculated by fitting the solubility expression, eq. (11), to experimental data of Wonders and Paul (15). The best fit to the experimental data was achieved with the parameters $\sigma o = 7.33 cm^3 (STP)/cm^3 (polymer) \cdot atm$ and $\alpha = 0.161 \; cm^3 (polymer)/cm^3 (STP)$. As can be seen in Fig. 1, eq. (11) describes the experimental data over the entire pressure range. The algorithm used to fit eq. (11) to the experimental data is described elsewhere (11).

The broken line in Fig. 1 was calculated following a non-linear least-squares fit of the dual-mode expression for the equilibrium gas concentration [eq. (14) in the preceding chapter] to the experimental points, with $k_D = 0.758 \; cm^3 (STP)/cm^3 (poly-mer) \cdot atm$, $C'_H = 14.58 \; cm^3 (STP)/cm^3 (polymer)$, and $b = 0.2835 \; atm^{-1}$ (15).

Both the matrix-model and the dual-model represent the experimental data satisfactory (Fig. 1). After modeling sorption measurements in several gas-polymer systems we have observed no systematic differences between the mathematical descriptions of the two models.

B. Transport of Gases in Glassy Polymers

1. Steady-State Transport

Solution of Fick's first law under the boundary conditions of a steady state permeation experiment yields an expression for the permeability constant, P (27),

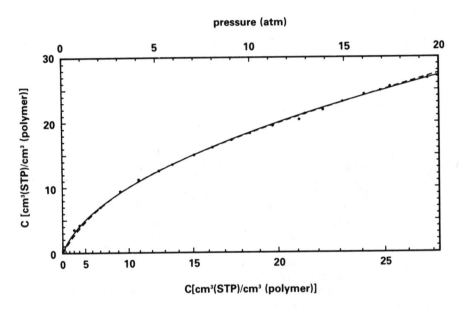

Figure 1. Sorption isotherm at 35 °C for CO_2 in polycarbonate conditioned by exposure to 20 atm CO_2. The experimental data are from Ref. 15. The curves, based on the matrix model (solid line) and the dual-mode model (broken line), are calculated using the parameters given in the text.

$$P = \bar{D} \, \sigma \tag{13}$$

When σ is given by eq. (4) and $\bar{D}$ by eq. (9) we have,

$$P = Do \; \sigma o \; \frac{1 + \beta C}{1 + \alpha C} \tag{14}$$

The solid line in Fig. 2 shows the permeability-pressure curve of CO_2 in polycarbonate calculated by fitting the simplified permeability expression, eq. (14), to experimental data of Wonders and Paul (15). The data fitting procedure, described elsewhere (11), gives $Do = 1.09 \times 10^{-8}$ cm^2/sec and $\beta = 0.065$ cm^3(polymer)/cm^3(STP). Fig. 2 shows a good agreement between the experimental data and eq. (14) over the entire pressure range.

Wonders and Paul (15) report that a nonlinear least-squares fit of the dual-mode expression [eq. (16) in the preceding chapter] to the permeability versus pressure data, for CO_2 in polycarbonate, gives D_D and D_H values of 4.78×10^{-8} and 7.11×10^{-9} cm^2/sec, respectively. The broken curve in Fig. 2 was calculated from the dual-mode sorption coefficients of Fig. 1 and the values of the diffusion coefficients given above.

Comparing the curves in Fig. 2 shows that representing the permeability versus pressure data by either model provides a satisfactory fit to the data over the pressure range of 1 to 20 atm. However, at pressures less than 1 atm. the two models differ in their prediction regarding the behavior of the permeability-pressure curve [Fig. 2]. While the matrix model predicts a strong apparent pressure dependence of the permeability in this range (solid line), the dual-mode model predicts only a weak dependence (broken line).

2. Transient Transport

Transient-transport measurements are a powerful tool for evaluating the validity of any sorption-transport model. The ability of a model to predict diffusion time lags is a test for its validity, as all the parameters are fixed by the equilibrium sorption and steady state transport, and because the time lag depends on the specific form of the concentration and diffusion gradients developed during the transient-state experiments.

Frisch has developed explicit expressions for the time lag in diffusion of a gas across a planar membrane when the diffusion coefficient is concentration dependent (29). Using his generalized method of solution,

$$\theta = \frac{\ell^2}{6Do} \cdot \frac{10 + 25 \, \beta C + 16(\beta C)^2}{10 \, (1 + \beta C)^3} \tag{15}$$

when $\bar{D}$ is given by eq. (9).

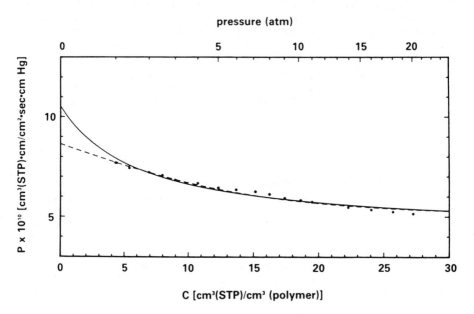

Figure 2. Permeability of CO_2 in conditioned polycarbonate at 35 °C. The experimental data are from Ref. 15. The solid curve is the calculated permeability based on the matrix model and the broken curve is calculated from the dual-mode model. Parameters used in the curves are given in the text.

Fig. 3 shows the calculated time lags (solid line) and the experimental time lags reported by Wonders and Paul (15) for CO_2 in polycarbonate. The time lags predicted by eq. (15) are in excellent agreement with the experimental results at both low and high pressures. The broken line in Fig. 3 results from calculations using the dual-mode model [eq. (17) in the previous chapter].

Using the dual-mode parameter values determined from sorption and permeation experiments, calculated time lags agree with the experimental data only at gas pressures above 5 atm. At lower pressures, dual-mode time lags are appreciably shorter than the observed ones, whereas time lags calculated from the matrix model by eq. (15) agree with the experimental data over the entire pressure range.

III. THE MATRIX AND DUAL-MODE MODELS AS PHYSICAL MODELS

The matrix model of gas sorption and transport ascribes the apparent concentration dependence of the structural and dynamic properties of the polymer to gas-polymer interactions. These interactions affect the interchain potential energy of the polymer, thus affecting the sorption and transport character-istics of the matrix. The matrix model is consistent with experimental evidence that the presence of gases causes changes in the structure, dynamics, glass-transition temperature, and viscoelastic relaxations of glassy polymers (7). The formalism developed here describes gas sorption and transport results and may also prove useful in quantifying these other changes.

In contrast, the dual-mode sorption and transport model assumes that the diffusant gas does not interact with the polymer (2, 4). In the dual-mode model the concentration dependence of the sorption and transport characteristics arises from the pressure-dependent distribution of the sorbed gas between Lang-muir sites and Henry's law dissolution. Experimental results presented in the preceding chapter (7) are inconsistent with the assumptions and the physical interpretations implicit in the dual-mode model. Although the dual-mode model provides a fairly adequate mathematical description of the phenomenological sorp-tion and transport of gases in glassy polymers (2, 4), the physical assumption of the model regarding no penetrant-polymer interactions and the existence of two distinct populations of sorbed gas molecules seems to be invalid in light of the experimental data (7).

A model for sorption and transport of gases in polymers should serve two purposes. First, the model should convey an understanding of the molecular interactions which are responsible for the macroscopic processes, and second, the model should

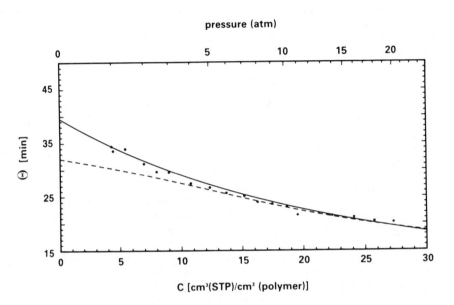

Figure 3. Time lag for diffusion of CO_2 at 35 °C in a 4.9 mil thick polycarbonate film conditioned by prior exposure to CO_2. The data are from Ref. 15. Calculated time lags based on the matrix model (solid line) and the dual-mode model (broken line) use parameters determined from fitting the sorption and permeation data.

enable correlative predictions of the macroscopic processes. Different mathematical expressions can be used for the modeling of the macroscopic sorption and transport phenomena. The dual-mode model proved to be extremely useful in this area (2). However, it is just as important to gain insight into the molecular processes in the gas-polymer matrix. It is mainly for this reason that we have introduced the matrix model and promoted its use in modeling macroscopic sorption and transport of gases in polymers. We believe that continued refinement of the formalism of the matrix model will lead to better understanding of gas-polymer systems.

Literature Cited

1. Barrer, R. M. J. Phys. Chem. 1957 61, 178, and references therein.
2. Paul, D. R. Ber. Bunsenges Phys. Chem. 1979, 83, 294, and references therein.
3. Crank, J. J. Polym. Sci. 1953, 11, 151.
4. Paul, D. R.; Koros, W. J. J. Polym. Sci., Polym. Phys. Ed. 1976, 14, 675.
5. Koros, W. J.; Paul, D. R.; Rocha, A. A. J. Polym. Sci., Polym.Phys. Ed. 1976, 14, 687.
6. Chern, R. T.; Koros, W. J.; Sanders, E. S.; Chen, S. H.; Hopfenberg, H. B., Chapter in this book.
7. Raucher, D.; Sefcik, M.D., Chapter in this book, and references therein.
8. Sefcik, M. D.; Schaefer, J.; May, F. L.; Raucher, D.; Dub, S. M. J. Polym. Sci., Polym. Phys. Ed. in press.
9. Sefcik, M. D.; Schaefer, J. J. Polym. Sci., Polym. Phys. Ed. in press.
10. Sefcik, M. D.; Raucher, D. J. Polym. Sci., Polym. Phys. Ed. in press.
11. Sefcik, M. D.; Raucher, D. J. Polym. Sci., Polym. Phys. Ed. in press.
12. Gee, G. Quart. Rev. Chem. Soc. 1947, 1, 265.
13. Pace, R. J.; Datyner, A. J. Polym. Sci., Polym. Phys. Ed. 1979, 17, 437.
14. Prigogine, I. "The Molecular Theory of Solutions"; North-Holland Publishing Co.: Amsterdam, 1957; Ch. 6.
15. Wonders, A. G.; Paul, D. R. J. Membrane Sci. 1979, 5, 63.
16. Assink, R. A. J. Polym. Sci., Polym. Phys. Ed. 1975, 13, 1665.
17. Zupančič, I.; Lahajnar, G.; Blink, R.; Reneker, D. H.; Peterlin, A. J. Polym. Sci., Polym. Phys. Ed. 1978, 16, 1339.
18. Schaefer, J.; Sefcik, M. D.; Stejskal, E. O.; McKay, R. A. Macromolecules in press.
19. Sefcik, M. D.; Andrady, A. L.; Schaefer, J.; Desa, E. Polym. Prepr., Am. Chem. Soc., Div. Polym. Chem. in press.

124INDUSTRIAL GAS SEPARATIONS

20. Assink, R. A. J. Polym. Sci. A-2, 1974, 12.
21. Wang, W. C.; Kramer, E. J. Bull. Am. Phys. Soc. 1981, 26, 367.
22. Wang, W. C.; Sachse, W.; Kramer, E. J. J. Polym. Sci. Polym. Phys. Ed., 1982, 20, 1371.
23. Chow, T. S. Macromolecules 1980, 13, 362.
24. Koros, W. J.; Paul, D. R.; Huvard, G. S. Polymer 1979, 20, 956.
25. Barrie, J. A.; Williams, M. J. L.; Munday, K. Polym. Eng. Sci. 1980, 20, 20.
26. Ranade, G.; Stannett, V.; Koros, W. J. J. Appl. Polym. Sci. 1980, 25, 2179.
27. Crank, J. "Mathematics of Diffusion", 2nd Ed.: Oxford Univ., 1975.
28. Ferry, J.D. "Viscoelastic Properties of Polymers," Wiley & Sons: New York, 1980, Ch. 15.
29. Frisch, H. L. J. Phys. Chem. 1957, 61, 93.

RECEIVED January 12, 1983

Membrane Gas Separations for Chemical Processes and Energy Applications

WILLIAM J. SCHELL

Separex Corporation, Spectrum Separations Division, Anaheim, CA 92806

C. DOUGLAS HOUSTON

Separex Corporation, DGH, Inc. Division, Tyler, TX 75703

Membranes have been used commercially for many years for water desalination by the reverse osmosis process. These membranes consist of a microporous substructure of cellulose acetate and a thin layer of dense cellulose acetate (active layer) on the upper surface, cast onto a supporting cloth for added mechanical strength. The active layer serves as the separating barrier, and due to its thinness, provides very high transport rates. It has been determined that these membranes, when dried, are also suitable for gas separation. As certain gases permeate more rapidly than others a gas mixture of two or more gases of varying permeability may be separated into two streams, one enriched in the more permeable components and the other enriched in the less permeable components. The membrane system to be described consists of spiral-wound elements connected in series and contained within pressure vessels. A rubber U-cup attached to the element serves to seal the element with the inner diameter of the pressure vessel, thereby forcing the feed gas to flow through the element. The pressure vessels usually contain six elements each and are mounted in racks on a skid. Applications that will be discussed include recovery of hydrogen from refinery process streams and tail gases, carbon dioxide removal and sweetening of natural gas, dehydration of natural gas on offshore platforms and production of carbon dioxide for enhanced oil recovery. The advantages of membrane separation over conventional processes typically include greatly reduced capital costs, lower energy consumption, smaller size and weight, lower installation costs due to its modular design, and simplified operation.

0097–6156/83/0223–0125$06.00/0
© 1983 American Chemical Society

It has been recognized for many years that nonporous polymer films exhibit a higher permeability toward some gases than towards others. As early as 1831, investigations were reported on the phenomenon of enrichment of air with rubber membranes (Ref. 1); however, not until 1950 had the practical possibility of this and other gas separations with permselective membranes been seriously studied. Weller and Steiner in their classic papers, demonstrated the feasibility of separating oxygen from air and described practical processes for separation of hydrogen and helium from methane (Ref. 2,3). Although their results were highly valuable in the development of the science of membrane separation, the calculated membrane area requirements for industrial processes were enormous, due to the very low permeation rates of the gases through these dense films.

The technical breakthrough in the application of membranes to gas separation came with the development of a process for preparing cellulose acetate in a state which retains its permselective characteristics but at greatly increased permeation rates (Ref. 4,5). These flat-sheet cellulose acetate membranes, which were orignally developed for reverse-osmosis water desalination (Ref. 6), are prepared from a solution of the polymer which is cast on a supporting cloth, partially dried then set or gelled in a water bath. At this state the membranes are heated in water to improve their selectivity characteristics, followed by drying with a solvent-exchange technique. These membranes have vastly increased permeation rates, while retaining the permselective characteristics of films of cellulose acetate, due to the formation of a thin, dense layer on the air-dried surface of the membrane. This so-called "active" layer has characteristics similar to those of cellulose acetate films but with a thickness of the order of 0.1 micrometer (μm) or less, whereas the total membrane thickness may range from approximately 75 to 125 μm (see Figure 1). The major portion of the membrane is an open-pore sponge-like support structure through which the gases flow without restriction. The permeability and selectivity characteristics of these asymmetric membranes are functions of casting solution composition, film casting conditions and post-treatment, and are relatively independent of total membrane thickness.

Methods were developed later to incorporate this asymmetric membrane structure for gas separation in a hollow fiber configuration rather than the flat-sheet (Ref. 7). Hollow fibers have a greater packing density (membrane area per packaging volume) than flat sheets, but typically have lower permeation rates. The mechanism for gas separation is independent of membrane configuration, however, and is based on the principle that certain gases permeate more rapidly than others. This is due to a combination of diffusion and solubility differences, whereby a gas mixture of two or more gases of varying permeability may be separated into two streams, one enriched in the more permeable components and the other enriched in the less permeable components.

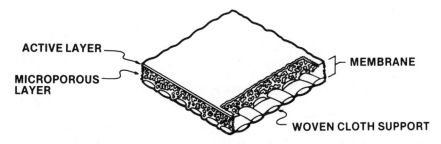

Figure 1. Cross section of asymmetric cellulose acetate membrane.

128   INDUSTRIAL GAS SEPARATIONS

Membrane Element Configuration

In order for membranes to be used in a commercial separation
system they must be packaged in a manner that supports the membrane and facilitates handling of the two product gas streams.
These packages are generally referred to as elements or bundles.
The most common types of membrane elements in use today include
the spiral-wound, hollow fiber, tubular, and plate and frame configurations. The systems currently being marketed for gas separation are of the spiral-wound type, such as the SEPAREX   and
Delsep   processes, and the hollow-fiber type such as the Prism
separator and the Cynara Company process.
Spiral-wound elements, as shown in Figure 2, consist primarily of one or more membrane "leaves," each leaf containing two membrane layers separated by a rigid, porous, fluid-conductive material known as the "permeate channel spacer." The permeate channel
spacer facilitates the flow of the "permeate", an end product of
the separation. Another channel spacer known as the "high pressure channel spacer" separates one membrane leaf from another and
facilitates the flow of the high pressure stream through the element. The membrane leaves are wound around a perforated hollow
tube, known as the "permeate tube", through which the permeate is
removed. The membrane leaves are sealed with an adhesive on three
sides to separate the feed gas from the permeate gas, while the
fourth side is open to the permeate tube.
The operation of the spiral-wound element can best be explained by means of an example. In order to separate carbon dioxide and/or hydrogen sulfide (acid gases) from a mixture of these
acid gases and hydrocarbon gases, the gaseous mixture enters the
pressure tube at high pressure and is introduced into the element
via the "high pressure channel spacer." The more permeable acid
gases rapidly pass through the membrane into the "permeate channel
spacer" where they are concentrated as a low pressure gas stream.
This low pressure acid gas stream flows through the element in the
"permeate channel spacer" and is continuously enriched by additional acid gas entering from other sections of the membrane.
When the low pressure acid gas stream reaches the permeate tube at
the center of the element, the acid gas "permeate" is removed.
The high pressure "residual gas" mixture remains in the high pressure channel spacer, losing more and more of its acid gas and
being enriched in hydrocarbon gas as it flows through the element,
and exits at the opposite end of the element.
The membrane system consists of membrane elements connected
in series and contained within pressure tubes as shown in Figure
3. A rubber U-cup attached to the element serves to seal the element with the inner diameter of the pressure tube, thereby forcing
the feed gas to flow through the element. The pressure tubes
usually contain six elements each and are mounted in racks on a
skid. Commercial size elements are typically 8 inches in diameter
by 40 inches long and contain from 150 to 275 square feet of
membrane area.

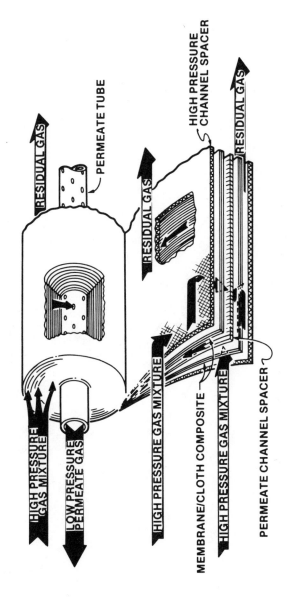

Figure 2.   Spiral-wound element construction.

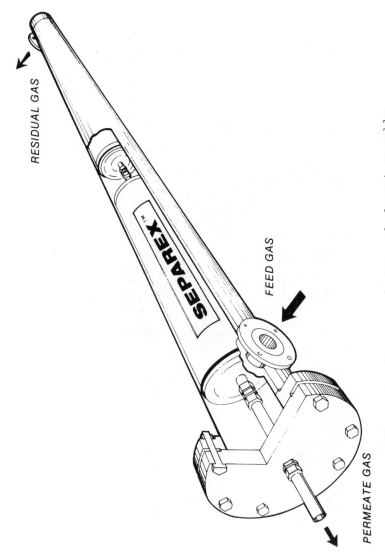

Figure 3.   Separex spiral-wound element assembly.

Typical operating parameters for the SEPAREX   process are a
temperature range of 32°F to 140°F, pressure differentials up to
1200 psig and feed flow rates from 50 MSCFD to more than 100
MMSCFD. The cellulose acetate membrane system is capable of processing mixtures containing a wide range of concentrations of hydrogen, carbon dioxide, hydrogen sulfide, water vapor, hydrocarbons and oxygen. Minor amounts of aromatic hydrocarbons, chlori-
nated hydrocarbons, olefins and heavy hydrocarbons do not appear
to effect the membrane performance. Elements have actually been
operated on liquid hydrocarbons without any detrimental effect.

Basic Equations Affecting Separation

The steady state mass flux (J_i) of component i through a
homogeneous film of uniform thickness separating two gaseous
phases is given by Fick's "First Law" of diffusion:

$$J_i = D_i \ dC_i/dx = \text{constant} \qquad 1)$$

where

D_i = local diffusivity (cm^2/sec)

C_i = local concentration of component i

x = the distance through the film

This relationship can be simplifed when the gases do not chemi-
cally associate with each other and when the gases are sparingly
soluble in the membrane material. In such cases, the diffusivity
of the permeating gas is constant through the film and the solubility of the gas at the membrane surface is essentially directly
proportional to its partial pressure in the gas phase adjacent to
that surface, i.e., Henry's Law applies:

$$C_i = k_i P_i \qquad 2)$$

where k_i is the solubility parameter and P_i is the partial
pressure.

If we let superscript I refer to the high pressure of the upstream side of the membrane and superscript II refer to the low
pressure or downstream side of the membrane, Equation 1, after
integration between C_i^I at x = 0 and C_i^{II} at x = ℓ, becomes:

$$J_i = \frac{k_i D_i \ (P_i^I - P_i^{II})}{\ell}$$

The product $k_i D_i$ is termed the permeability coefficient of the
membrane for component i. This coefficient is independent of membrane thickness and pressure differential and the frequently used
units are, cc(STP) - cm/cm^2-sec-cm Hg. Another parameter of

interest is the permeation rate, $\bar{P}_i$, defined as $k_i D_i/\ell$, which is a measured characteristic of a given membrane, with units cc(STP) /cm^2-sec-cm Hg or $SCFH/ft^2 \cdot 100$ psi. The total pressures, P^I and P^{II}, are given by the sums, $P_i^I + P_j^I$ and $P_i^{II} + P_j^{II}$, respectively. The ratio $\bar{P}_i/\bar{P}_j$ is defined as the ideal separation factor for component i with respect to component j in the membrane and is written $\alpha i/j$.

From the previous discussion it can be seen that, if component i is the more permeable, increasing P_i^I, either by increasing the total pressure or the concentration of component i, will result in a higher membrane permeability rate. In addition, higher values for $\alpha i/j$ result in greater efficiency in gas separation.

It can be shown that the composition of the permeate gas, P_i^{II}, for permeation of a binary gas mixture is given by the following quadratic equation:

$$\frac{P_i^{II}}{(P^{II} - P_i^{II})} = \frac{\alpha i/j \quad (P_i^I - P_i^{II})}{(P^I - P^{II}) - (P_i^I - P_i^{II})}$$

Parameters Affecting The Process

The separation efficiency for a given membrane with a particular binary gas mixture will be dependent mainly upon three factors: gas composition, the pressure ratio between feed and permeate gas, and the sepration factor for the two components. A higher separation factor gives a more selective membrane, resulting in a greater separation efficiency. This parameter is a function of the membrane material and is determined by the individual gas permeation rates.

The gas composition and pressure ratio are interrelated in terms of their effect on separation efficiency. Their interaction is pronounced when the feed gas has a low concentration of the more permeable gas, with the result that the effective separation is small at low pressure ratios, irrespective of the $\alpha i/j$ value of the membrane. This behavior is shown in Figure 4. This phenomenon is due to the fact that the partial pressure of the more permeable component on the permeate (low pressure) side cannot exceed its partial pressure on the feed (high pressure) side. For high concentration gases the pressure ratio has only a small influence, as shown in Figure 5. Hence, if compression energy is an important factor in the economics of a particular separation, lower pressure ratios may be used with little or no loss in separation efficiency. By contrast (see Figure 4), the pressure ratio is a very important consideration in achieving efficient separation of low concentration gas mixtures.

Another important factor in membrane gas separation is the pressure differential, $P^I - P^{II}$; the greater this difference the less membrane area required. The membrane area is exactly proportional to the inverse of the pressure differential, i.e., Area = constant $(P^I - P^{II})^{-1}$. The effective separation factor is

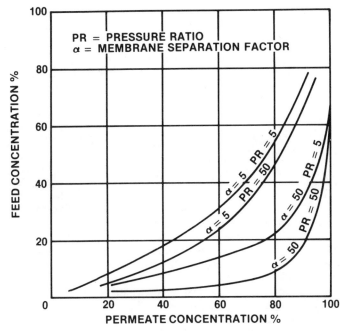

Figure 4.   Effect of feed composition on permeate enrichment.

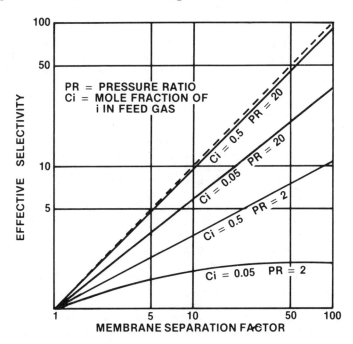

Figure 5.   Effect of pressure ratio and membrane separation factor on effective selectivity.

unaffected by the pressure differential as long as the pressure
ratio is unchanged.
    Other system variables that will have an effect on the sepa-
ration process are temperature and relative humidity of the gas.
Increasing the temperature raises most permeabilities by about 10
to 15% per 10°C and has little effect on separation factors. The
effect of relative humidity is variable depending upon the membrane used. High relative humidities, greater than 95%, are generally detrimental due to membrane plasticization. Contamination
with liquid water has been found to dramatically reduce membrane
performance for cellulose acetate.
    The flow dynamics of the permeate stream through the product
channel spacer material, or through the fiber bore in the case of
hollow fiber systems, may also adversely affect the system performance. If the permeate flow is sufficiently high, a back pressure will be observed in the permeate stream. It can be seen from
Figures 4 and 5 that this would result in a lower enrichment due
to a lowering of the effective pressure ratio. It has been found
that in spiral-wound elements the permeate flow behaves as if it
were under laminar conditions. Gas viscosity is therefore the
determining factor in comparing the relative back pressure of
various gases flowing at the same rates. This effect may be
essentially eliminated by proper construction of the membrane
element.

Cellulose Acetate Membrane Data

    Membranes manufactured by Spectrum Separations, Inc., a sub-
sidiary of SEPAREX CORPORATION, are of the cellulose acetate type.
They are similar to those made for reverse osmosis except they
must be dried for gas separation use. A proprietary process is
used to accomplish this so that the membrane does not collapse
and lose its asymmetric character upon removal of the water.
    The permeation rates are measured at various pressures for
the gases of interest in order to characterize the membranes.
Permation rates for water vapor, helium, hydrogen, hydrogen sulfide and carbon dioxide are the highest with N_2, CH_4, CO and C_2H_6
being the lowest. The high rate for H_2O, H_2S and CO_2 is due to
the high solubility of these gases in the membrane. Gas permeation and selectivity data for SEPAREX^{TM} spiral-wound elements
are shown in Table I. It may be noted that the oxygen/nitrogen
selectivity is of an order of magnitude less than the other gas
couples, due to the very small difference in molecular size and
solubility of O_2 and N_2. This low separation factor makes oxygen
enrichment the most difficult to accomplish, both technically and
economically, of the major gas separations of interest. The selectivity and permeation rates achieved in finished elements are
somewhat lower than the membrane due to boundary layer problems,
permeate back pressure, problems with imperfect seals, defects in
the membrane and contaminants found in gas mixture in the field.

Table I ELEMENT PERFORMANCE

Gas Couple	Separation Factor, α
H_2O (Vapor)/CH_4	200 - 400
H_2/CH_4	45 - 55
CO_2/CH_4	20 - 30
H_2S/C_3H_8	75 - 110
He/CH_4	60 - 85
O_2/N_2	4 - 5

Permeation Rate at 500 psig, SCFH

Gas	4-in. Dia.	8-in. Dia.
He	3,300	14,000
H_2	2,800	12,000
CO_2	1,400	6,000
O_2	200	850

Applications

There are many commercial applications for membrane gas separations, some of which are currently being marketed, and others being tested.

Several field test studies have been undertaken utilizing the SEPAREX process in a 2-in. diameter element size. Due to the modular configuration of membrane systems, a full size system can be directly designed from the test results with a small pilot plant. Although the flow rates for a pilot unit are considerably lower than might be encountered in a full-size system, all process parameters such as product purities, pressure drop, product recoveries, optimum pressure and temperature, membrane area required and series/parallel arrangement of the elements can be directly determined.

Acid Gas Separation from Methane. Removal of hydrogen sulfide (H_2S) and carbon dioxide (CO_2) from natural gas is an ideal application for membranes in that H_2S and CO_2 permeate through the membrane approximately 30 times faster than membrane, enabling a high recovery of the acid gases without significant loss of pressure in the methane pipeline product gas (Ref. 5,8). The membrane system would be considered a bulk removal process in this case and has the greatest economic advantage over conventional technology when the acid gas content is over 10%.

The most obvious place for membrane systems in this application is in the retrofitting of existing sour gas processing plants. This would increase the capacity and reduce the energy load of the existing system or eliminate the need for expanding the existing plant when wellhead pressure loss or increased acid gas content occurs.

Another application for membrane systems related to sour gas processing is the recovery of carbon dioxide for use as a miscible flood for enchanced oil recovery (EOR) from depleted oil fields (Ref. 9). It has been found that CO_2 at concentrations above 90% will solubilize oil absorbed in the substrata, allowing for secondary and tertiary recovery. There are many gas wells containing about 40% to 80% CO_2 that could be processed with a membrane system, providing a high purity CO_2 for EOR and a fuel gas by-product. In EOR production, CO_2 injected into the formation eventually appears with light hydrocarbon gases associated with the crude oil. A membrane system can again be used to recover CO_2 for reinjection and recover light hydrocarbons for sale as pipeline gas. For example, a typical associated gas contains 70-85% CO_2 at 100°F and 40 psig (Ref. 10). If such gas is sent through a two-stage SEPAREX   process at 300 psig, CO_2 would be recovered in the permeate stream at a concentration of 95% and a recovery of 98%, along with a residual high pressure light hydrocarbon pipeline gas containing 3% or less CO_2.

A CO_2-CH_4 methane process gas stream, similar to a typical high CO_2 natural gas has been under test by SEPAREX for CO_2 removal in a 2-in. diameter element pilot plant since September 1981. The feed gas contains 30% CO_2 and is delivered to the membrane test unit at 250-450 psig under ambient temperature conditions. The objective of the system is to reduce the CO_2 level of the methane to less than 3.5%. The membrane system consists of 5 pressure tubes in series, each tube containing three 40-in. long elements. The gas is conditioned to maintain it at a minimum of 20°F above the dewpoint. The system was operated at a variety of flow rates, pressures, recoveries and temperatures. Selected data are presented in Figures 6 through 8.

A plot of the feed and residual gas flow rates as a function of stage cut is shown in Figure 6. Stage cut is defined as the fraction of feed gas that permeates through the membrane for a given number of elements (single stage). In this case a stage consists of fifteen elements in series. It can be seen that at a feed flow rate of about 180 SCFH all of the gas would permeate the membrane, resulting in no enrichment. This is further demonstrated in Figure 7, where the permeate gas composition at zero residual gas flow is equal to the feed composition.

By combining Figures 6 and 7 it is possible to relate stage cut to the residual and permeate gas stream compositions. These calculations are useful in determining recoveries and in designing series/parallel flow arrangements in large systems. The dashed portions of the curves in the low flow rate region of Figure 7 represents predicted CO_2 compositions under uniform flow conditions, whereas the solid lines represent actual obtained data. The deviation between predicted and actual performance is most likely due to feed gas channeling in the elements or the development of stagnant flow regions as a result of the low pressure difference across the high pressure side (feed to residual) of the elements.

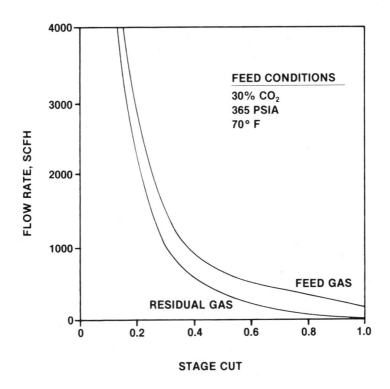

Figure 6. Stage cut as function of flow rate.

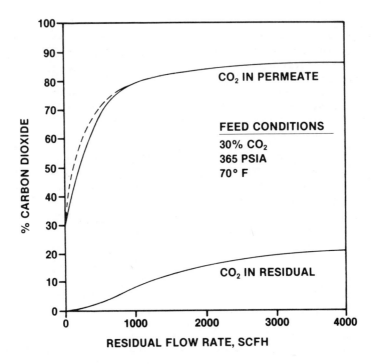

Figure 7. Effect of flow rate on enrichment.

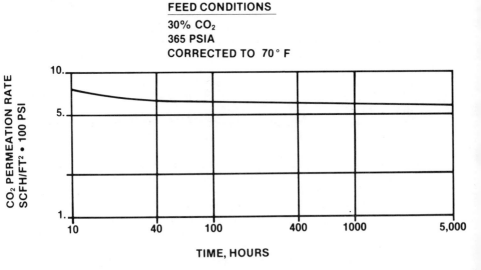

Figure 8. Time-dependent flux decline (log/log).

Under the particular feed conditions and element arrangement for this test, the criteria of less than 3.5% CO_2 in the residual stream could not be reached without experiencing some loss in separation efficiency. This goal could be achieved, however, by operating at higher flow rates with more elements in series. The conclusion arrived at from these data, therefore, is that there is a critical minimum flow rate for given feed gas conditions and element array. In order to maximize system performance this parameter must be taken into consideration when designing a full size system.

Time-dependent flux decline is plotted in Figure 8. The CO_2 permeation rate, in units of $SCFH/ft^2$.100 psi, was calculated from performance data for the fifteen elements in series. This rate corresponds to that which would be obtained for pure CO_2 under similar partial pressure differences. After an initial (40 hrs) nonlinear flux decline, the CO_2 rate experienced a 10% loss over the next 1,000 hours (40 days). As this flux decline is a log/log relationship, only 6% more would be lost over the next three years, the minimum expected element lifetime. Membrane systems can be easily designed to adapt to a changing permeation rate by adjusting the feed flow rates and/or pressure, by allowing for addition of more elements in series after a period of time, and by using a progressive element replacement schedule.

Gas Dehydration. It has been found that water vapor permeates cellulose acetate membranes at a rate approximately 500 times that of methane (Ref. 2). This exceptionally high selectivity for water vapor make cellulose acetate membrane systems attractive for dehydration of hydrocarbon gas streams to pipeline specifications on either a pure gas stream or while simultaneously removing contaminating acid gases. For these applications the small size, low weight and low maintenance of the SEPAREX^{TM} system is particularly advantageous for offshore installations.

Liquid water is detrimental to the performance of the membrane, however, so that the feed gas is delivered to the membrane system at less than 90% relative humidity. This can be accomplished by either heating the saturated gas 10-20°F or by dropping the pressure slightly.

The previously described pilot plant test was operated with gases containing water vapor for some period of time during the test program. The feed gas was saturated at 450 psig in a range of temperatures from 80-95°F, corresponding to water contents of 62-100 lbs/MMSCF. The feed gas pressure was dropped from 450 psig prior to entering the membrane unit. Two pressures were studied; 250 psig (57% relative humidity) and 350 psig (78% relative humidity). The membrane unit was operated for 2 months under these conditions with no deterioration of the carbon dioxide permeation rate or change in membrane selectivity. The feed gas under all conditions was dehydrated to a dew point below -55°F (<5 lbs

water/MMSCF), or greater than 97% water removal. The water vapor
measuring device used was not suitable below -55°F so that an
accurate measurement of the dew point was not obtainable. The
measurement was further complicated by the presence of traces of
heavy hydrocarbons interfering with the water dew point. It was
concluded from this study, however, that Spectrum Separations'
cellulose acetate membranes are suitable for dehydration even in
the presence of high CO_2 concentrations.

Hydrogen Recovery. Hydrogen is consumed in petroleum refining, in the manufacture of ammonia and methanol, for hydro-
genation of fats and oils and in the electronics industry, at an
annual rate of approximately 7 million tons (Ref. 11). In many
of these processes hydrogen is lost as an off-gas that could be
purified and recycled to the point of use. Most of these off-
gases are mixtures of hydrogen and carbon monoxide, methane and/or
nitrogen and offer an excellent application for membrane systems
due to a high hydrogen permeation rate and the high selectivity of
hydrogen relative to these gases.

    An additional application for membranes might be realized if
coal gasification and coal liquefaction processes continue to
develop to a commercial level. The membrane system could be uti-
lized to recover bydrogen from the hydrogasifier off-gas and be
recycled to the gasifier or sold as a chemical feedstock. In
other coal gasification schemes, the product gas could be upgraded
or separated into a hydrogen product and a fuel gas product.

    SEPAREX delivered a hydrogen recovery system, utilizing
4-in. diameter spiral-wound elements, in early 1982. The system
will recover hydrogen from the off-gas of a unit utilizing UOP's
Butamer[R] process in a LPG processing complex.

    In UOP's Butamer process normal butane is catalytically
isomerized to produce high-purity iso-butane. N-butane is charged
into a fixed-bed vapor phase reactor and hydrogen is added with
small amounts of an organic chloride compound. The unit yields
iso-butane and a hydrogen stream contaminated with hydrocarbons
and traces of HCl. The hydrogen stream is fractionated, leaving
an off-gas containing about 70% hydrogen. This gas, which normally is sent to fuel, is processed with the SEPAREX membrane
system to recover the hydrogen for recycle to feed make-up. A
simplified process schematic is shown in Figure 9.

    The SEPAREX system will recover over 90% of the hydrogen at
a purity of 96+% for recycle, while increasing the heating value
of the fuel gas from ~550 BTU/SCF to ~950 BTU/SCF. The projected
flow rates and gas purities for the membrane separation are shown
in Table II. Under the bone-dry feed conditions the cellulose
acetate membrane is not affected by HCl. Special materials of
construction and adhesives have been used in the fabrication of
the spiral-wound elements to ensure their resistance to HCl in the
gas streams.

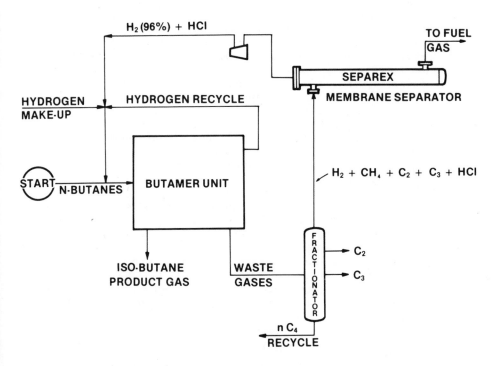

Figure 9.   Hydrogen recovery schematic.

Table II   HYDROGEN RECOVERY FROM
                         A UOP's BUTAMER[R] PROCESS

	Feed Gas	Residual Gas	Permeate Gas
Pressure, psia	265	240	15
Flow Rate, MSCFH	47.8	16.8	31.0
Temperature, °F	110	100	100
Mol% H_2	68.9	17.8	96.4
Mol% CH_4	23.7	63.0	2.6
Mol% C_2+	6.8	19.0	0.2
Mol% HCl	0.6	0.2	0.8

Oxygen Enrichment of Air. The U.S. produces approximately
15 million tons of oxygen from air each year (Ref. 12). Most of
this production is accomplished cryogenically, and in large plants
the cost is very low. Uses of oxygen in quantities of less than
10 tons per day, however, constitute a substantial share of this
market. It is thought that membrane systems could economically
enrich air at these smaller plant sizes.

    One area of particular interest is in secondary sewage treat-
ment utilizing oxygen instead of air (Ref. 13). There are
approximately 5,000 air-treatment sewage plants in this country
that each consume 5 tons of atmospheric oxygen per day. Many of
these plants suffer seasonal overloads (for example in farming
communities) and many others are reaching capacity and will require new construction for expansion. The use of enriched oxygen
in place of air could increase the capacity of these existing
plants without the addition of space-consuming conventional treatment processes.

    Five major U.S. chemical firms are now producing oxygenation
systems for sewage wastewater treatment, in what the industry
feels could become oxygen's largest market (Ref. 12). Membrane
systems could economically compete with conventional technologies
in the smaller plants.

Literature Cited

1. Mitchell, J. V. J. Roy. Inst. 1831, 2, 101, 307.
2. Weller, S.; Steiner, W. A. J. Appl. Physics 1950, 21, 279.
3. Weller, S.; Steiner, W. A. Chem. Eng. Prog. 1950, 46, (11),
    585.
4. Gantzel, P. K.; Merten, U. Ind. Eng. Chem. 1970, 9, (2), 331.
5. Schell, W. J.; Lawrence, R. W.; King, W. M. Membrane Applica-
    tions to Coal Conversion Processes, Final Report to the
    Energy Research and Development Administration, Report No.
    FE-2000-4, 1976.
6. Reid, C. E.; Breton, E. J. J. Appl. Poly. Sci. 1959, 1, 133.
7. Gardner, R. J.; Crane, R. A.; Hannon, J. F. Chem. Eng. Prog.
    October 1977, 76.
8. Cooley, T. E.; Coady, A. B. U. S. Patent 4,130,403 1978.

9. Goddin, C. S. Comparison of Processes Treating Gases with High CO$_2$ Content, presented at the 61st Annual GPA Convention, March 1982.
10. Meissner, R. E. Hydrocarbon Processing, April 1980, 113.
11. Kelley, J. H.; Escher, W. J. D.; vanDeelen, W. Chem. Eng. Prog. January 1982, 58.
12. Fallwell, W. F. Chem. & Eng. News, July 15, 1974, 7.
13. Gross, R. W. Chem. Eng. Prog. October 1976, 51.

RECEIVED January 12, 1983

Things that adsorbed → adsorbates
Things that do adsorb + adsorbent

8

Gas-Adsorption Processes: State of the Art

GEORGE E. KELLER, II

Union Carbide Corporation, Ethylene Oxide and Glycol Division,
Technical Center, South Charleston, WV 25303

Gas-adsorption processes are used for a wide
variety of separations throughout the chemical and
other industries. But to effect various separations,
quite different process embodiments are necessary.
These differences are primarily concerned with the
method by which the adsorbent is regenerated. This
choice is influenced by the percentage of adsorbate
in the feed, ease of desorption and degree of
separation required. In this paper, we discuss
temperature-swing, inert-purge, displacement-purge,
pressure-swing, parametric-pumping, chromatographic
and other cycles; the separations for which they
are best suited; and their limitations. Commercial
examples are given. Finally, predictions are made
as to future improvements in various technologies
as well as to what new separations will be amenable
to adsorption separation.

Gas-adsorption processes involve the selective concentration
(adsorption) of one or more components (adsorbates) of a gas (or
vapor) at the surface of a microporous solid (adsorbent). The
attractive forces causing the adsorption are generally weaker
than those of chemical bonds and are such that, by increasing
the temperature of the adsorbent or reducing an adsorbate's
partial pressure, the adsorbate can be desorbed. The desorption
step is quite important in the overall process. First,
desorption allows recovery of adsorbates in those separations
where they are valuable, and second, it permits reuse of the
adsorbent for further cycles.

This paper is concerned with a broad range of gas-adsorption
processes as practiced in the organic-chemical, petroleum,
natural-gas and allied industries. Emphasis will be on a
description of processes rather than a review of adsorber design

0097–6156/83/0223–0145$07.25/0

procedures, which have been covered in other publications (1-6),
or a discussion of traditional adsorbents, which have also been
covered elsewhere (7-15). The objectives of this paper are
(i) to review those situations in which adsorption can be an
economically viable process, (ii) to provide a description of
present-day adsorption technology, and (iii) to indicate the
likely directions that new adsorption technology and uses will
take.

Criteria for When to Use Gas Adsorption

Distillation and related vapor-liquid processes are by far
the most common means of separating homogeneous mixtures in the
industries mentioned above. The reasons for this are the basic
process simplicity and scalability of distillation (which helps
promote low investment per annual unit of feed) and the ability
to achieve many theoretical stages (which effects high degrees
of separation) in many systems. On the other hand, distillation
is fundamentally a high-energy-usage process. In most distillations the energy usage is at least one and frequently two
orders of magnitude greater than the minimum thermodynamic work
of separation.

We can use the following as rough criteria to decide
whether gas adsorption might be a viable alternative for
distillation for a given separation.

1. The relative volatility between the key components to
    be separated is in the order of 1.2 to 1.5 or less.
2. The bulk of the feed is a relatively low-value, more-
    volatile product, and the product of interest is in
    relatively low concentration (about 10 weight percent
    or less). In such situations large reflux ratios can
    be required, resulting in high energy costs.
3. One set of components whose boiling range overlaps
    that of a second set of chemically or geometrically
    dissimilar components must be separated from the
    second set. To make such a separation, even though
    various relative volatilities may be reasonably large,
    several distillation columns would be required.
4. Separation by distillation requires cryogenic
    operation or pressures above about 20 atmospheres.

Gas adsorption should be considered for a given separation
when one or more of the above criteria apply to distillation
and when a suitable adsorbent exists. In general a suitable
adsorbent is one which shows adequate selectivity (greater than
two for the key components) and capacity, can be regenerated
easily and causes no damage to the products by promoting
byproduct-forming reactions. It is also usually desirable that
the adsorbent selectively adsorb the component(s) in lower
concentrations in the feed to minimize the amount of adsorbent
required per unit of feed processed.

Separations and Processes

   Commercial gas-adsorption processes (see Table I) can be
divided into bulk separations, in which about 10 weight percent
or more of a stream must be adsorbed, and purifications, in
which usually considerably less than 10 weight percent of a
stream must be adsorbed. Such a differentiation is desirable
to make because in general different process cycles are used
for the different categories, as will be discussed later.
Below, four basic process cycles and two combinations are
described in their simplest forms. In other sections, recent
uses and modifications of these cycles, as well as other new
process cycles, are described.

   Temperature-Swing Cycle. In this cycle, a stream containing
a small amount of an adsorbate is passed through the adsorbent
bed at a relatively low temperature. After equilibrium between
adsorbate in the feed and on the adsorbent is reached, the bed
temperature is raised to a higher value, and more feed is passed
through the bed. Desorption occurs and a new, lower equilibrium
loading is established. The net bed removal capacity, called
the delta loading, is the difference between these two loadings.
(This value is actually an upper limit, since equilibrium
loadings in practical operation are not attained.) When the
bed is subsequently cooled and feed is again passed through the
bed, purification will occur down to a concentration in
equilibrium with the residual loading on the adsorbent.
   The time required to heat, desorb and cool a bed is usually
in the range of a few hours to over a day. Because during this
long regeneration time the bed is not productively separating,
temperature-swing processes are used almost exclusively to
remove small concentrations of adsorbates from feeds. Only in
such situations can the on-stream time be a significant fraction
of the total cycle time of the process.

   Inert-Purge Cycle. In this cycle, the adsorbate, instead
of being removed by temperature increase, is removed by passing
a non-adsorbing gas containing no adsorbate through the bed.
This has the effect of lowering the partial pressure of the
adsorbate around the particles, and desorption occurs. If
enough purge gas is passed through the bed, the adsorbate will
be completely removed, and the maximum delta loading will equal
the equilibrium loading. As in the temperature-swing cycle and
all others, actual delta loadings will be less than the
equilibrium delta loading. During the subsequent adsorption
step, removal of adsorbate from the feed would be essentially
complete until the adsorbent becomes nearly fully loaded and
breakthrough occurs.

TABLE I

REPRESENTATIVE COMMERCIAL GAS-ADSORPTION SEPARATIONS

Separation*	Adsorbent
I. Gas Bulk Separations	
Normal Paraffins/Isoparaffins, Aromatics	Zeolite
N_2/O_2	Zeolite
O_2/N_2	Carbon Molecular Sieve
CO, CH_4, CO_2, N_2, A, NH_3/H_2	Zeolite, Activated C
Acetone/Vent Streams	Activated C
C_2H_4/Vent Streams	Activated C
Separation of Perfume Components	---
II. Gas Purification	
H_2O/Olefin-Containing Cracked Gas, Natural Gas, Air, Synthesis Gas, Etc.	Silica, Alumina, Zeolite
CO_2/C_2H_4, Natural Gas, Etc.	Zeolite
Organics/Vent Streams	Activated C, Others
Sulfur Compounds/Natural Gas, Hydrogen, Liquified Petroleum Gas (LPG), Etc.	Zeolite
Solvents/Air	Activated C
Odors/Air	Activated C
NO_x/N_2	Zeolite
SO_2/Vent Streams	Zeolite
Hg/Chlor-Alkali Cell Gas Effluent	Zeolite

---

*Adsorbates are listed first.
  which is adsorbed

Displacement-Purge Cycle. This cycle, which is somewhat
similar to the previous one, differs from it in that a gas or
vapor which adsorbs about as strongly as the adsorbate is used
to remove the adsorbate (see Figure 1). Removal is thus
facilitated both by adsorbate partial-pressure reduction in the
fluid around the particles and by competitive adsorption of the
displacement medium. As with the inert-purge cycle, the
maximum delta loading is the equilibrium loading.
    The use of the two different types of purge fluids causes
two major differences in the processes. First, since the
displacement-purge fluid is actually adsorbed, it is present
when the adsorption part of the cycle begins and therefore
contaminates the less-adsorbed product. (In the inert-purge
cycle this contamination is usually much smaller.) In practical
terms this means that the displacement-purge fluid must be
recovered from both product streams. Second, since the heat of
adsorption of the displacement-purge fluid is approximately
equal to that of the adsorbate, as the two exchange on the
adsorbent, the heat generated (or consumed) is virtually zero,
and the adsorbent's temperature remains unchanged. When an
inert-purge gas is used, a temperature rise occurs in the
region where adsorption occurs, and this temperature rise can
severely limit the adsorption capacity of the bed in some cases.

    Pressure-Swing Cycle. In a gas-adsorption cycle, the
partial pressure of an adsorbate can be reduced by reducing the
total pressure of the gas. This change can be used to effect a
desorption. The lower the total pressure of the desorbing
step, the greater the maximum delta loading will be.
    The time required to load, depressurize, desorb and
repressurize a bed is usually a few minutes and can in some
cases be only a few seconds. Thus, even though practical delta
loadings are substantially less than the maximum delta loadings
to minimize thermal-gradient problems, the very short cycle
times make a pressure-swing cycle quite attractive for bulk-gas
separations.

    Combined Cycles. Often a temperature-swing cycle is
combined with an inert-purge to further facilitate desorption.
Several possibilities are shown in Figure 2. The inert-purge
stream can be a fraction of the less-adsorbed product, or a
separate purge stream can be used. If the feed is at super-
atmospheric pressure, many times the desorption step will be
carried out at atmospheric pressure.
    In pressure-swing processes, it is quite common to use a
fraction of the less-adsorbed gas product as a low-pressure
purge gas, as shown in Figure 2. Often the purge flow is in
the opposite direction from the feed flow. Rules for the
minimum fraction of less-adsorbed gas product for displacing

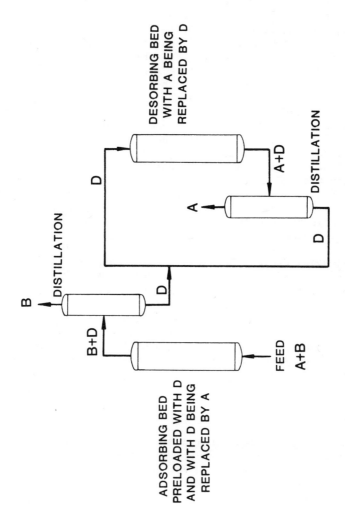

Figure 1. Displacement-purge cycle. A, adsorbate; B, less-adsorbed
component; and D, displacement agent.

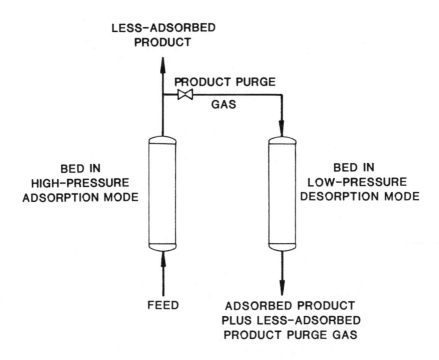

Figure 2.   Combined pressure-swing, non-adsorbed-product-purge
cycle (pressure-swing adsorption).

the adsorbate have been given by Skarstrom (16). This process
has been called heatless fractionation or, more commonly,
pressure-swing adsorption (PSA). The latter name will be used
in this paper.

Gas Bulk-Separation Processes

Gas bulk-separation processes consist mostly of pressure-
swing adsorption (PSA) variations and displacement-purge and
inert-purge processes. Recently, however, a chromatographic
cycle has been commercialized. Below, these cycles will be
discussed.

PSA. As mentioned earlier, PSA processes use a fraction
of the less-adsorbed product to aid in purging the adsorbate
from the adsorbent. Thus an inevitable feature of present-day
PSA processes is the loss of part of the less-adsorbed product
to the purge stream. This has the additional implication that
the purge stream cannot be of high purity, and as a result PSA
is limited at present to those applications where only one pure
product is desired: oxygen and nitrogen (but not both
simultaneously) from air, hydrogen recovery in cases in which
complete recovery is not required, and chemical process
purge-stream treatment in which the less-adsorbed product is
removed from the process and the adsorbate-enriched stream is
recycled to the process.
       Most often, PSA processes use three or more beds in
parallel. The advantages are (i) energy savings, which accrue
from using gas from a high-pressure bed to partially
repressurize a low-pressure bed following desorption, and
(ii) higher product recoveries, which accrue from using a
combination of purging steps in the cycle.
       Recently two widely divergent process variations of PSA
have been commercialized by Union Carbide Corporation. The
first of these is called POLYBED PSA, which is used for hydrogen
recovery (17-21). Plants with capacities of up to about 1.2 x
10^6 NM^3/day of hydrogen have been built. Hydrogen recoveries
of 86 percent vs 70-75 percent for other PSA processes have
been demonstrated, as well as purities of 99.999 mole percent
(20). POLYBED PSA involves the use of five or more beds, with
extensive gas interchanges and pressure equalizations between
the beds. One process embodiment (17) is shown in Figure 3.
       Thus POLYBED PSA can be thought of as a process which aims
for maximum recovery of the less-adsorbed component in very
high purity at the expense of a complex flowsheet. Its growing
commercial acceptance suggests that it competes quite
successfully with cryogenic separation in many cases.
       The second process, called pressure-swing parametric
pumping (PSPP), was developed to minimize process complexity

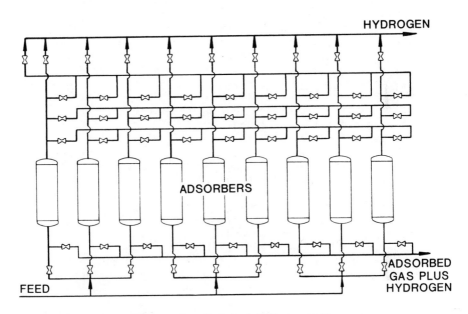

Figure 3. Polybed PSA process.

and investment at the expense of product recovery. The name
parametric pumping was coined by Wilhelm (22), who described an
adsorption-based separation process involving reversing flows.
When the flow is in one direction a parameter, such as
temperature, which influences adsorptivity, is at one value,
while the parameter is changed to another value when the flow
is in the opposite direction. Such a process will create a
separation between components with different adsorptivities.
Chen (23) has correctly pointed out that PSA processes
constitute a subset of parametric pumping, in which pressure is
the parameter used to influence adsorptivity.

Because of its highly unusual nature, PSPP will be
described in some detail. PSPP exists in both single-bed
(24-26) and multiple-bed (27) embodiments. Schematic diagrams
of both are shown in Figure 4. The single-bed process will be
described first. The bed can vary from about 0.3 to 1.3 or
more meters in length and contains adsorbent particles with
diameters in the range of 177 to 420 um. Feed gas under
pressure is supplied in pulses of up to about a second in length
from a compressor and a surge tank. The pulse is controlled by
a solenoid valve and a timer. During this feed pulse the
exhaust solenoid valve is closed. Following the feed pulse
both valves at the feed end are closed for about 0.5 to three
seconds; this period is called the delay. Next the solenoid
valve on the exhaust or purge line opens for a period of about
five to twenty seconds. Since the pressure in this line is
maintained below that in the feed line, reverse flow of gas
from the bed occurs. And finally another delay period with
both solenoid valves closed is used in some cases. Its length
is generally less than one second.

However, while pressures and flow directions are
fluctuating substantially at the feed end of the bed, a
continuous flow emerges from the product end and through a
small surge tank. This flow is enriched in less-adsorbed
components; in the case of air separation, the product consists
typically of 90 to 95 percent oxygen, with the balance argon
and a small amount of nitrogen. Conversely the exhaust stream
is somewhat depleted in oxygen and argon.

The net effect of this process is to produce, using a
single adsorbing bed and two surge tanks, a constant flow of
less-adsorbed product - primarily oxygen in this case - from a
constant flow of air from a compressor.

The multiple-bed process uses essentially the same range
of bed lengths and adsorbent particle sizes. The time cycles
of the three beds can be sequenced in such a way that compressed
feed is always flowing to the process. This reduces the size
of the feed surge tank or completely eliminates the need for it.
In general the fraction of the overall cycle devoted to the
feed pulse in the multiple-bed process is much greater - about

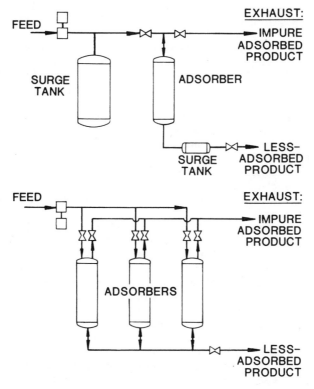

Figure 4. One-bed and three-bed pressure-swing parametric pumps.

one-third – than the fraction in the single-bed process – in the order of one-tenth. The multiple-bed process usually produces somewhat higher less-adsorbed product pressures and slightly higher less-adsorbed product recoveries compared to the single-bed process.

The most unusual feature of PSPP is the continuously changing axial pressure profile. Whereas all other processes operate with virtually a constant pressure through the bed at any given time, the fast cycle of PSPP and the flow resistance caused by the use of very small adsorbent particles create substantial pressure drops in the bed, as shown in Figure 5. These pressure profiles are highly critical to the performance of the process. For example, the unusual profile during the exhaust part of the cycle, which shows a maximum pressure somewhere inside the bed, simultaneously permits purging of part of the bed while less-adsorbed products are continually produced at the opposite end of the bed.

PSPP has been commercialized for the production of oxygen and for the recovery and recycle of ethylene and a small amount of chlorocarbons (the adsorbed stream) to an ethylene-chlorination process while purging nitrogen (the less-adsorbed component) from the process.

As mentioned before, POLYBED PSA and PSPP represent quite different approaches to improving the basic PSA process. Table II shows the pluses and minuses of these approaches compared to a basic PSA process.

A new PSA process has been developed by Bergbau Forschung GmbH for the production of nitrogen from air (28,29). The process uses a unique carbon-based molecular sieve as the adsorbent. The adsorbent's micro-geometry is such that, even though the adsorption isotherms of oxygen and nitrogen are almost identical, oxygen diffuses many times more rapidly into the pores. This creates an oxygen-depleted gas phase until both gases equilibrate, which requires over an hour. By operating the process with a time cycle considerably less than the equilibrium time, the adsorbent becomes oxygen-selective and produces a high-nitrogen product. Nitrogen purities of up to 99.9 percent can be produced. The process, depicted in Figure 6, uses a vacuum desorption step to facilitate production of high-purity nitrogen.

Nitrogen production using carbon molecular sieves is the only known commercial process using differences in intraparticle diffusivity, rather than inherent adsorbent selectivity or selective molecular exclusion, as the basis for the separation.

Another method of using PSA to obtain nitrogen from air has recently been revealed by Toray Industries, Inc. (30). The standard PSA process has been modified to produce an impure (33 percent oxygen) less-adsorbed gas and a relatively pure (up to 99.9 percent) purge stream of nitrogen. A diagram is shown

TABLE II

COMPARISON OF THE PERFORMANCES OF POLYBED   PSA
AND PRESSURE—SWING PARAMETRIC PUMPING
WITH BASIC PSA

	Polybed PSA	Pressure—Swing Parametric Pumping
Compression Cost	Less	Same or Greater
Productivity	About the Same	Much Greater
Degree of Separation	Greater	Same or Less
Process Complexity	Greater	Less
Adaptability to Large Flows	Greater	Less
Cycle Time	About the Same (Several Minutes)	Much Less (Several Seconds)
Non-Adsorbed Product Pressure	Same	Lower

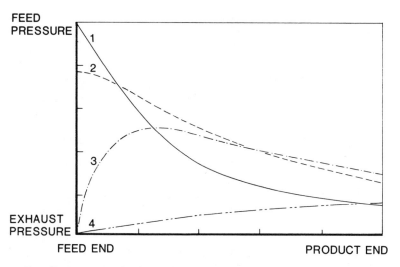

Figure 5. Pressure profiles in an adsorbent bed producing 90 mole percent oxygen. 1, middle of feed; 2, middle of delay; 3, early in exhaust; and 4, late in exhaust.

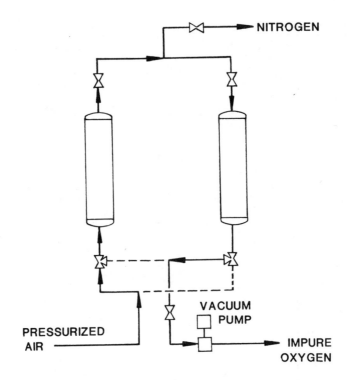

Figure 6.   Bergbau Forschung nitrogen-production process.

in Figure 7. Dried air is passed through the adsorber at super-
atmospheric pressure; nitrogen is preferentially adsorbed. Part
of the nitrogen product from previous cycles is then passed
into the bed to desorb small amounts of oxygen, after which the
bed is reduced to atmospheric pressure to desorb nitrogen.
Finally more nitrogen is recovered by vacuum desorption. The
adsorbent used in this process is apparently the same – zeolite
molecular sieves – as that used in oxygen PSA processes.
    This process has been demonstrated in pilot-scale. Its
commercial status is unknown.

    Displacement–Purge and Inert–Purge Cycles. By far the
most widespread embodiment of these cycles is the separation of
normal and iso–paraffins in a variety of petroleum fractions.
These fractions in general contain several carbon numbers and
can include molecules from about C_5 to about C_{18}. All of these
separations use 5A molecular sieve as the adsorbent. Its
0.5 nm pore diameter is such that normal paraffins can enter
but iso–paraffins are excluded; this constitutes the basis for
separation.
    A recent review (13) lists six different commercial vapor-
phase processes. These processes use either displacement–purge
or inert–purge cycles. The former are used when higher-
molecular–weight fractions are being processed. In such cases,
the adsorbates are held so tightly by the adsorbent that the
use of an adsorbing displacement agent is necessary to effect
desorption. In turn, however, two distillation columns are
required to recover the displacement fluid from the product
streams. When lower–molecular–weight fractions are being
processed, the simpler inert–purge cycles suffice. With these
cycles, coolers followed by vapor–liquid separators can be used
to recover products from the purge gas.
    The normal/iso–paraffin separation process used most often
is Union Carbide's IsoSiv process. IsoSiv units with feed
capacities up to 3600 metric tons per day have been built. In
the mid–1970s IsoSiv was combined with the Hysomer paraffin-
isomerization process developed by Shell Research B.V. (31,32);
this combination is called the Total Isomerization Process (TIP)
and can produce a completely iso–paraffin product from a normal
or mixed paraffin feed. A schematic diagram is shown in Figure
8. Hydrogen, which is required in the isomerization step, also
serves as the purge gas for desorbing the normal paraffins,
which are recycled to the Hysomer unit. The process is carried
out in the vapor phase, and utility requirements are considerably
below those of distillation (32). Product from a TIP unit has
a Research Octane Number of 88 to 92, compared to a RON of 79 to
82 for the product from a Hysomer unit alone.
    The close–coupling of reaction and separation units for
driving equilibrium reactions to complete conversion, though

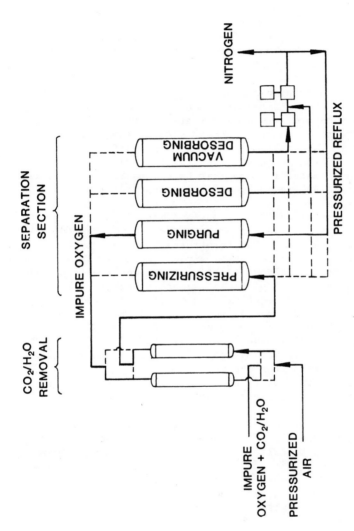

Figure 7. Toray reverse PSA nitrogen-production process.

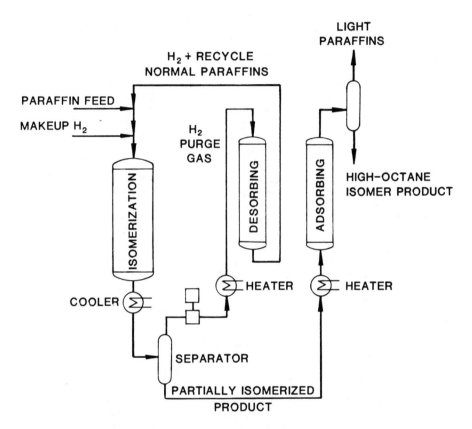

Figure 8.   Total isomerization process.

not new, is probably an underutilized concept. Adsorption,
which can sometimes separate materials easily which distillation
can separate only with difficulty, represents an extra degree
of freedom in conceiving of such processes.

Another isomer-separation process which has been announced
is called OlefinSiv (13). This process separates iso-butene
from normal butenes in a manner similar to IsoSiv.

Chromatography. The laboratory analytical technique of
gas chromatography has been an appealing model for a commercial
separation process for many years. In spite of numerous
attempts at scale-up, only recently has this effort apparently
been successful. The Societe Nationale Elf Aquitaine and
Societe de Recherches Techniques et Industrielles (SRTI) have
announced a process for separating mixtures difficult to
separate by distillation (33-35). Hydrogen is used as the
carrier gas. The Elf-SRTI process is presently commercially
separating 100 metric tons per year of perfume ingredients.
More recently Elf Aquitaine has announced a 100,000 metric tons
per year demonstration unit for separating C_4-C_{10} normal and
iso-paraffins (35). A schematic diagram is shown in Figure 9.
This process uses several beds in parallel to produce a constant
flow of products.

As mentioned earlier, the use of a non-condensable,
non-adsorbing carrier gas simplifies the task of recovery of
products compared to that task in displacement-purge adsorption.
At the same time, the scale-up of this process means that the
effects of separation-inhibiting thermal gradients accompanying
adsorption and desorption have been dealt with successfully.

Gas-Purification Processes

Compared to adsorption's use in bulk-gas separations, its
use in gas purifications is much more frequent (see Table I and
references 13, 36 and 37), and the technology is, for the most
part, more conventional. Temperature-swing adsorption, often
combined with inert-purge stripping, is by far the most common
process used. Two or more fixed beds operated in parallel,
typically with one adsorbing and one or more regenerating,
constitute the standard flowsheet.

Temperature-swing cycles are inherently energy-intensive
per unit of material adsorbed; regeneration energy usage is
usually several times the heat of condensation of the adsorbate.
However, if the amount of adsorbate per unit of feed is small,
the regeneration energy usage per unit of feed can reduce to a
reasonable value. A second advantage is that a combined
temperature-swing/inert-purge cycle can provide very high
degrees of desorption during regeneration, leading to very high
degrees of adsorbate removal from the feed. Removals down to

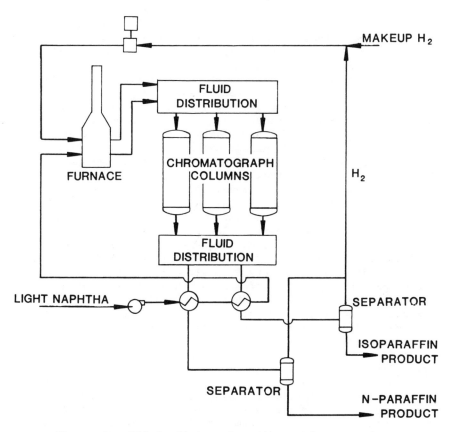

Figure 9.   Elf Aquitaine chromatographic separator.

one volume part per million or less of adsorbate in the
less-adsorbed product are not uncommon.

The development of a very hard, microspherical activated
carbon called bead activated carbon (BAC), by Kureha Chemical
Co., Ltd. of Japan has made possible the commercialization of a
new fluidized-bed/moving-bed process (38). This process was
called GASTAK by Taiyo Kaken Co., Ltd. (now a part of Kureha)
and is licensed by Union Carbide Corporation for sale in the
United States and Canada under the name PURASIV HR (39). A
diagram is shown in Figure 10. This process is used primarily
for removal of solvent vapors from air streams. Feed gas is
contacted countercurrently by fluidized adsorbent on trays
similar to those in distillation columns. The loaded adsorbent
then drops into a heated, moving-bed regeneration zone where
the adsorbate is desorbed and recovered. Stripped adsorbent is
then gas-lifted to the top tray in the upper section. Adsorbate
reductions in the less-adsorbed gas can range from 90 to well
over 99 percent (40). The advantages claimed for this process
over fixed-bed processes include energy reductions of 70 to 75
percent, improved quality of recovered adsorbate, mechanical
simplicity and compact size.

The PURASIV HR process is not the first circulating-
adsorbent process to be commercialized. As early as the 1930s
moving-bed processes were proposed. About 1950, Union Oil
Company developed the Hypersorption moving-bed process for
various light-hydrocarbon separations (41,42), and in the
mid-1960s, Cortaulds, Ltd. developed a fluidized-bed process
for carbon disulfide recovery from air (43). These processes
were plagued by attrition and loss of the activated carbon
adsorbent, however, and were shut down. BAC apparently
eliminates that problem, raising the possibility of applying
fluidized-bed and moving-bed adsorption technology to a number
of other separations.

Future Directions

As stated earlier, distillation and related vapor-liquid
operations are by far the most common separation processes in
the organic-chemical, petroleum and allied industries. It is
unlikely that adsorption will ever rival distillation in
frequency of use, but adsorption's share of the separation task
should grow substantially. The purpose of this section is to
indicate the likely areas of growth in existing and new
applications, as well as the technological innovations which
would foster this growth.

Gas Bulk Separations. Recovery of oxygen and nitrogen by
various pressure-swing processes will continue to grow. Oxygen
from these processes is already being produced for waste-
treatment plants, for welding shops and for in-home care of

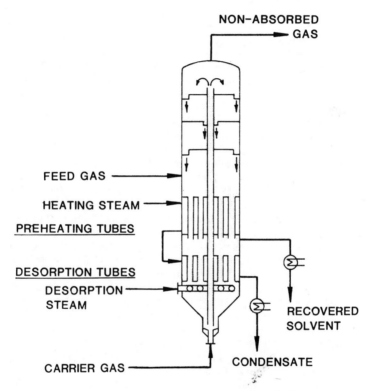

Figure 10. PURASIV HR fluidized-bed/moving-bed process.

lung-disease patients. Recently it was announced that units to
supply oxygen on board aircraft will be built (44). Nitrogen
will be produced for inerting of storage tanks, purging of
process lines and vessels, creating modified atmospheres to
prolong storage life in food-storage areas, etc.

It can also be predicted that, as further process
improvements are made, PSA oxygen will compete successfully
with cryogenically-produced oxygen in quantities considerably
above 10 to 20 metric tons per day.

PSA hydrogen recovery will also compete more favorably with
cryogenic alternatives in the future. Streams of 50 to 100
thousand cubic meters per hour will be processed at pressures up
to about 70 atmospheres. In addition, as hydrogen becomes more
valuable in chemical processes, high-purity hydrogen via PSA
will be used more frequently to minimize vent-stream losses.
Conversely impure hydrogen in other vent streams will be a
candidate for recovery and cleanup via PSA.

But pressure-swing processes will not compete with
distillation for many gas separations requiring the production
of two nearly pure products (examples: ethane/ethylene, removal
of inerts from natural gas) without a technological breakthrough.
That breakthrough would be the ability to perform such
separations in a relatively mechanically simple, i.e., low-
investment, fashion, and with energy requirements less than
those of comparable distillations. Though these are formidable
goals, they should not be considered impossible to attain.

Displacement-purge cycles will not find many new uses,
primarily because of their inherent complexity (see Figure 1).
This complexity prevents these cycles from competing with
distillation from an investment standpoint except in cases
where key-component relative volatilities are well below 1.5.
Inert-purge cycles will find additional uses if the problem of
dissipating or coping with the heat of adsorption, which can
cause large temperature rises during adsorption and thereby
limit adsorption capacity, can be solved.

Moving-bed and fluidized-bed processes should be
reinvestigated for economic viability, given the existence of
Kureha's BAC and the possiblity of producing other adsorbents
in a tough, microspheroidal form. Also, the recent work of
Szirmay (45) certainly gives impetus to moving-bed processes by
showing that quite sharp separations can be made between key
components in very short lengths of bed. A major problem of
such processes is the same as that of the fixed-bed, inert-purge
cycles: the large temperature rise accompanying adsorption,
such as was found in the Hypersorption process (41).

Finally, there may be a few opportunities for close-
coupling of reaction and adsorption systems to overcome
thermodynamic-equilibrium limitations or to enhance selectivity
by operating with low conversions per pass. Reaction types

include hydrogenation/dehydrogenation, isomerization and
esterification/deesterification, among others, with boiling
points of the species low enough to be amenable to vapor-phase
processing.

Gas Purifications. Few if any major process innovations
are likely to occur in this area. Fixed-bed, temperature-swing
processes will continue to dominate, with some extension of
PURASIV HR technology into new areas. Applications of
adsorption will grow, however. A major area for growth is the
recovery of small amounts of organics from chemical-process,
storage-tank and other gaseous vents, as well as from solvent-
painting, purging and cleaning operations. This area will grow
in direct relation to the concern over air-pollution problems.
Other major-use areas – gas dehydration, removal of sulfur
compounds and carbon dioxide, and various specialty separations
(37) – will grow in proportion to the growth in the industries
they serve.

Acknowledgment

The author gratefully acknowledges helpful discussions
with and sharing of information by Messrs. R. A. Anderson,
R. T. Cassidy, M. S. Clancy, R. L. Drahushuk, and M. F. Symoniak,
all of Union Carbide Corporation.

Literature Cited

1. Lukchis, G. M. Chemical Engineering June 11, 1973, 111.
2. Lukchis, G. M. Chemical Engineering July 9, 1973, 83.
3. Lukchis, G. M. Chemical Engineering August 6, 1973, 83.
4. Vermeulen, T.; Klein, G.; Hiester, N. K. "Adsorption and
   Ion Exchange," in "Chemical Engineers' Handbook," Perry,
   R. H.; Chilton, C. H., Eds., 5th edition, McGraw-Hill,
   New York, 1973.
5. Vermeulen, T. "Adsorptive Separation," in "Kirk-Othmer
   Encyclopedia of Chemical Technology," 3rd edition, vol. 1,
   Wiley-Interscience, New York, 1978.
6. Kovach, J. L. "Gas-Phase Adsorption," in "Handbook of
   Separation Techniques for Chemical Engineers," Schweitzer,
   P. A., Ed., McGraw-Hill, New York, 1979.
7. Deitz, V. R. "Bibliography of Solid Adsorbents," N.B.S.
   Circular 566, National Bureau of Standards, Washington, DC,
   1956.
8. "Adsorption Handbook," Pittsburgh Chemical Co., Pittsburgh,
   PA, 1961.
9. Young, D. M.; Crowell, A. D. "Physical Adsorption of
   Gases," Butterworths, London, 1962.
10. Ross, S.; Olivier, J. P. "On Physical Adsorption," Wiley-
    Interscience, New York, 1964.

168   INDUSTRIAL GAS SEPARATIONS

11. Breck, D. W. "Zeolite Molecular Sieves," Wiley-Interscience,
    New York, 1974.
12. Ponec, V.; Knor, Z.; Cerny, S. "Adsorption on Solids,"
    Butterworths, London, 1974.
13. Chi, C. W.; Cummings, W. P. "Adsorptive Separation (Gases),"
    in "Kirk-Othmer Encyclopedia of Chemical Technology," 3rd
    edition, vol. 1, Wiley-Interscience, New York, 1978.
14. Flanigen, E. M. "Molecular Sieve Technology – The First
    Twenty-Five Years," in "Proceedings of the Fifth
    International Conference on Zeolites, Naples, Italy, 2–6,
    June, 1980," Heyden, London, 1980.
15. Breck, D. W.; Anderson, R. A. "Molecular Sieve Zeolites,"
    in "Kirk-Othmer Encyclopedia of Chemical Technology," 3rd
    edition, vol. 15, Wiley-Interscience, New York, 1981.
16. Skarstrom, C. W. "Heatless Fractionation of Gases over
    Solid Adsorbents," in "Recent Developments in Separation
    Science," Li, N. N., Ed., vol. 2, CRC Press, Cleveland, OH,
    1972.
17. Fuderer, A.; Rudelstorfer, E. "Selective Adsorption
    Process," United States Patent 3,986,849, October 19, 1976.
18. Heck, J. L. Hydrocarbon Processing January, 1978, 175.
19. Corr, F.; Dropp, F.; Rudelstorfer, E. Hydrocarbon
    Processing March, 1979, 119.
20. Cassidy, R. T. "Polybed Pressure-Swing Adsorption Hydrogen
    Processing," in "Adsorption and Ion Exchange With Synthetic
    Zeolites," Flank, W. H., Ed., American Chemical Society
    Symposium Series 135, 1980.
21. Heck, J. L. Oil and Gas Journal February 11, 1980, 122.
22. Wilhelm, R. H. Ind. Eng. Chem. Fund. 1966, 5, 141.
23. Chen, H. T. "Parametric Pumping," in "Handbook of
    Separation Techniques for Chemical Engineers," Schweitzer,
    P. A. (Ed.), McGraw-Hill, New York, 1979.
24. Jones, R. L.; Keller, G. E.; Wells, R. C. "Rapid Pressure-
    Swing Adsorption With High Enrichment Factor," United
    States Patent 4,194,892, March 25, 1980.
25. Keller, G. E.; Jones, R. L. "A New Process for Adsorption
    Separation of Gas Streams," in "Adsorption and Ion Exchange
    With Synthetic Zeolites," Flank, W. H., Ed., American
    Chemical Society Symposium Series 135, 1980.
26. Jones, R. L.; Keller, G. E. Journal of Separation Process
    Technology 1981, 2 (3), 17.
27. Earls, D. E.; Long, G. N. "Multiple Bed Rapid Pressure
    Swing Adsorption for Oxygen," United States Patent
    4,194,891, March 25, 1980.
28. Knoblauch, K. Chemical Engineering November 6, 1978, 87.
29. "Pure Nitrogen Generator," Gas Services International,
    Ltd., Enfield, Middlesex, England.
30. Miwa, K.; Inoue, T. Chemical Economy and Engineering Review
    1980, 12 (11), 40.

31. Symoniak, M. F.; Reber, R. A.; Victory, R. M. Hydrocarbon Processing May, 1973, 101.
32. Symoniak, M. F. Hydrocarbon Processing May, 1980, 110.
33. Bonmati, R. G., Chapelet-Letourneux, G., and Margulis, J. R. Chemical Engineering March 24, 1980, 70.
34. Anon. Analytical Chemistry 1980, 52, 481A.
35. Bernard, J. R.; Gourlia, J. P.; Guttierrez, M. J. Chemical Engineering May 18, 1981, 92.
36. Lee, M. N. Y. "Novel Separation With Molecular Sieves Adsorption," in "Recent Developments in Separation Science," Li, N. N., Ed., vol. 1, CRC Press, Cleveland, OH, 1972.
37. Anderson, R. A., "Molecular Sieve Adsorbent Applications State of the Art," American Chemical Society Symposium Series 40, 1977.
38. Sakaguchi, Y. Chemical Economy and Engineering Review 1976, 8 (12), 36.
39. Anon. Chemical Engineering August 29, 1977, 39.
40. "PURASIV HR for Hydrocarbon Recovery," Union Carbide Corporation, Danbury, CT.
41. Berg, C. Petroleum Refiner 1951, 30 (9), 241.
42. Treybal, R. E. "Mass Transfer Operations," McGraw-Hill, New York, 1955.
43. Avery, D. A.; Tracey, D. H. "The Application of Fluidized Beds of Activated Carbon to Solvent Recovery from Air or Gas Streams," Institution of Chemical Engineers Symposium Series, No. 30, 1968.
44. Anon. Science News 1981, 120, 346.
45. Szirmay, L. "Design Aspects of MAB (Moving Adsorbent Bed) Units for Continuous Chromatography-Like Sharp Separation," presented at the A.I.Ch.E. National Meeting, Orlando, FL, February 29, 1982.

RECEIVED January 25, 1983

Nonisothermal Gas Sorption Kinetics

SHIVAJI SIRCAR and RAVI KUMAR
Air Products and Chemicals, Inc., Allentown, PA 18105

Analytical equations for adsorbate uptake
and radial adsorbent temperature profiles
during a differential kinetic test are derived.
The model assumes that the mass transfer into
the adsorbent can be described by a linear
driving force model or the surface barrier
model. Heat transfer by Fourier conduction
inside the adsorbent mass in conjunction with
external film resistance is considered.
Experimental uptake data for sorption of
i-octane on 13X zeolite and n-pentane on 5A
zeolite were quantitatively described by the
model. The results show that internal thermal
resistance of the adsorbent mass plays a
significant role during the uptake for these
systems even though the adsorbent temperature
changes are small.
The model shows that the non-isothermal
uptake curve for an adsorbent mass which has
low effective thermal conductivity (k_e) is
identical in form to that of the isothermal
Fickian diffusion model for mass transport.
It is shown that k_e can be significantly low
for an assemblage of microparticles at low
pressure and high temperatures.
A parametric study of the effects of the
equilibrium and the transport properties of
the adsorption system on sorption kinetics is
carried out. Complex interactions between
these properties in determining the shape of
the uptake curve are observed.

0097–6156/83/0223–0171$07.00/0

The importance of adsorbent non-isothermality during
the measurement of sorption kinetics has been recognized in
recent years. Several mathematical models to describe the
non-isothermal sorption kinetics have been formulated [1-9].
Of particular interest are the models describing the uptake
during a differential sorption test because they provide
relatively simple analytical solutions for data analysis
[6-9]. These models assume that mass transfer can be
described by the Fickian diffusion model and heat transfer
from the solid is controlled by a film resistance outside
the adsorbent particle. Diffusion of adsorbed molecules
inside the adsorbent and gas diffusion in the interparticle
voids have been considered as the controlling mechanism for
mass transfer.

Comparisons of estimated diffusivity values on zeolites
from sorption uptake measurements and those obtained from
direct measurements by nuclear magnetic resonance field
gradient techniques have indicated large discrepancies
between the two for many systems [10]. In addition, the
former method has often resulted in an adsorbate diffusivity
directly proportional to the adsorbent crystal size [11].
This led some researchers to believe that the resistance to
mass transfer may be confined in a skin at the surface of
the adsorbent crystal or pellet (surface barrier) [10,11].
The isothermal surface barrier model, however, failed to
describe experimental uptake data quantitatively [10,12].

Measurement of radial temperature gradient inside the
adsorbent particle during the sorption test showed that the
internal resistance to heat transfer may not be negligible
in comparison with the external film [13-16].

In view of these observations, we propose a non-
isothermal sorption kinetics model with the following
assumptions:

(a) Sorbate mass transfer can be described by a linear
     driving force model (LDF) using a "lumped-up" mass
     transfer coefficient k_s [17].
(b) Fourier conduction inside the adsorbent mass in
     conjunction with an external film resistance describes
     the heat transfer from the adsorbent.

The LDF model is a realistic representation of the
system with a surface barrier. Otherwise, k_s can be treated
as an apparent mass transfer coefficient irrespective of the
true transport mechanism which can be directly used in the
design and optimization of adsorbers. This concept has been
successfully used to analyze column breakthrough data for
practical non-isothermal systems [18-20]. It substantially

simplifies the mathematical modelling of the behavior of adsorbent columns in comparison with the use of Fickian diffusion models [22,23].

Non-isothermal LDF Model

We assume that the adsorbent mass used in the kinetic test consists of a sphere of radius R. It may be composed of several microsize particles (such as zeolite crystals) bonded together as in a commercial zeolite bead or simply an assemblage of the microparticles. It may also be composed of a noncrystalline material such as gels or aluminas or activated carbons. The resistance to mass transfer may occur at the surface of the sphere or at the surface of each microparticle. The heat transfer inside the adsorbent mass is controlled by its effective thermal conductivity. Each microparticle is at a uniform temperature dependent on time and its position in the sphere.

We assume that the adsorbent mass is initially in equilibrium with the adsorbate at pressure P_0 and temperature T_0, and a differential step change in the gas phase pressure to P_∞ is applied at time $t = 0$. The pressure inside the adsorbent mass (or the microparticles), $P(t)$, increases with time, which is given by:

$$\frac{d\,P(t)}{dt} = k_s \, [P_\infty - P(t)] \tag{1}$$

k_s is the "lumped-up" mass transfer coefficient.

The heat balance for the adsorbent in the region $0 \leq r < R$ may be written as:

$$\rho c \left. \frac{\partial T}{\partial t} \right|_r = \frac{1}{r^2} \cdot \frac{\partial}{\partial r} [k_e r^2 \cdot \frac{\partial T}{\partial r}] + \rho q \left. \frac{\partial n}{\partial t} \right|_r \tag{2}$$

Where T is the temperature of the adsorbent mass at radius r and time t. n is the adsorbate loading per unit weight of the adsorbent at radius r and time t. q is the isosteric heat of adsorption. ρ, c and k_e are, respectively, the density, the heat capacity and the effective thermal conductivity of the adsorbent mass.

The adsorbent is at equilibrium under local conditions of P and T. Thus for a differential test where the changes in the adsorbent temperature and the adsorbate loading are small, one may write:

$$[n\,(t,r) - n_\infty] = a\,[P - P_\infty] + b\,[T(t,r) - T_0] \tag{3}$$

$$a = \frac{\partial n^*}{\partial P} \text{ at } T_o, P_\infty \tag{4}$$

$$b = \frac{\partial n^*}{\partial T} \text{ at } T_o, P_\infty \tag{5}$$

The asterick indicates that the quantities are evaluated under equilibrium conditions. n_o and n_∞ are, respectively, the initial and the final equilibrium adsorbate loadings during the test.

The boundary conditions for equations (1) and (2) are:

$$
\begin{aligned}
P(o) &= P_o \\
P(\infty) &= P_\infty \\
T(r,o) &= T_o \\
T(r,\infty) &= T_o
\end{aligned}
$$

$$\frac{\partial T}{\partial r}\bigg|_{r=0} = 0$$

$$\frac{\partial T}{\partial r}\bigg|_{r=R} = -\frac{h}{k_e} \cdot [T-T_o]$$

The last boundary condition accounts for the external heat transfer from the adsorbent mass. h is the effective external heat transfer coefficient. a, b, c, ρ, q, k_e, k_s and h can be assumed to be constants for a differential test.

The average adsorbate loading in the adsorbent mass, $\bar{n}(t)$, can be obtained by:

$$\bar{n}(t) = \frac{3}{R^3} \int_o^R n\,(r,t) \cdot r^2 \cdot dr \tag{6}$$

Equations (1) - (3) and (6) were simultaneously solved using the above boundary conditions by Laplace transformation and inversion by the method of residues to obtain the following analytical equations:

$$F = 1 - e^{-\eta\tau} + \frac{6\beta}{1-\beta} \quad \frac{\eta}{\eta - p_n^2} \quad \frac{s^2}{p_n^2\{p_n^2 - s(1-s)\}}$$

$$(e^{-p_n^2\tau} - e^{-\eta\tau}) \tag{7}$$

$$\theta = 2 \quad \frac{\sin p_n x}{(x) \sin p_n} \quad \frac{\eta}{\eta - p_n^2} \quad \frac{s}{\{p_n^2 - s(1-s)\}}$$

$$(e^{-p_n^2 \tau} - e^{-\eta \tau}) \qquad (8)$$

where

$$x = r/R$$

$$\tau = \frac{\alpha t}{R^2}$$

$$\beta = \frac{bq}{c}$$

$$s = \frac{h R}{k_e}$$

$$\alpha = \frac{k_e}{\rho c (1-\beta)} \qquad (9)$$

$$\eta = \frac{k_s R^2}{\alpha}$$

$$F = \frac{\bar{n} - n_o}{n_\infty - n_o}$$

$$\theta = \frac{T(x,t) - T_o}{T^* - T_o}$$

$$(T^* - T_o) = \frac{q(n_\infty - n_o)}{c(1 - \beta)}$$

p_n's are the positive roots of the transcedental equation:

$$p_n \cot p_n - 1 = -s \qquad n = 1,2,3,\ldots \qquad (10)$$

It can be shown from adsorption thermodynamics that:

$$\beta = - \frac{q^2 (n_o + \eta_\infty)/2}{c \, R_g T_o^2} \left. \frac{\partial \ell n \, n}{\partial \ell n \, P} \right|^*_{at \, T_o, (P_o + P_\infty)/2} \qquad (11)$$

a, q and β can be calculated from the equilibrium adsorption isotherms at various temperatures. R_g is the gas constant.
Equation (7) gives the dimensionless uptake, F, for the differential test in terms of dimensionelss time (τ).
Equation (8) describes the dimensionelss absorbent temperature θ as a function of dimensionless time and radius (x). The paratmeters of these two equations are dimensionless groups, η [reciprocal of modified Lewis number], s[Biot number] and the equilibrium parameter β.
Equations (7) and (8) have nine variables, Viz. a, q, b, c, ρ, R, k_s, and h. Of these variables, the equilibrium properties [a, q, β] can be evaluated by independent measurement of adsorption isotherms at two or more temperatures. The physical properties of the adsorbent mass [c, ρ, R] can also be independently measured. The remaining three transport properies [k_s, k_e, h] can then be evaluated from the experimental uptake data and the temperature-time profiles in the adsorbent at two different radial positions using equations (7) and (8). However, measurement of T(x,t) may be difficult in a differential test.
For moderately fast adsorption ($\eta \gg p_1^2$), it can be shown from equation (7) that a plot of $\ln [1-F(t)]$ against t yields a straight line at large t. Such behavior has been experimentally observed for many systems [7,8,24]. The slope (m) and intercept (I) at t = 0 of the linear curve are given by:

$$m = - \frac{\alpha p_1^2}{R^2} \qquad (12)$$

$$I = - \frac{6\beta}{1-\beta} \frac{\eta}{\eta - p_1^2} \frac{s^2}{p_1^2 \{p_1^2 - s(1-s)\}} \qquad (13)$$

where p_1 is the first root of equation (10)

$$p_1 \cot p_1 - 1 = - s \qquad (14)$$

Equation (7) shows that the slope (m_o) of $\ln[1-F(t)]$ against t at t = 0 is:

$$m_o = - \frac{\eta \alpha}{R^2 (1-\beta)} \qquad (15)$$

Equations (12)-(15) provide four independent equations relating the parameters η, p_1, α/R^2 and s. Thus a unique solution is possible. However, m_o may be difficult to measure for systems with relatively fast adsorption. In that case, the problem reduces to a single variable search to best fit the uptake curve using equations (12)-(14).

Special Cases

For isothermal adsorption [$q\rightarrow 0$], equation (7) reduces to

$$F = 1 - e^{-k_s t} \qquad (16)$$

For very rapid adsorption [$k_s\rightarrow\infty$], equations (7) and (8) become:

$$F = 1 + \frac{6\beta}{1-\beta} \sum \frac{s^2}{p_n^2\{p_n^2-s(1-s)\}} e^{-p_n^2 \tau} \qquad (17)$$

$$\theta = 2 \sum \frac{\sin p_n x}{(x)\sin p_n} \frac{s}{\{p_n^2 - s(1-s)\}} e^{-p_n^2 \tau} \qquad (18)$$

Equation (17) represents the general case of uptake under complete heat transfer control and equation (18) gives the corresponding adsorbent temperature profile. Furthermore, if k_e is large compared to h, (s << 1), equation (10) can be approximated as $p_n^2 \sim 3$ s and equations (17) and (18) reduce to

$$F = 1 + \frac{\beta}{1-\beta} e^{-\frac{3ht}{R\rho c(1-\beta)}} \qquad (19)$$

$$\theta = e^{-\frac{3ht}{R\rho c(1-\beta)}} \qquad (20)$$

Equations (19) and (20) correspond to the familiar case of uptake controlled by external heat transfer resistance derived earlier by King and Cassie [25]. Equation (20) shows that (T^*-T_o) is the maximum temperature rise in the adsorbent for this case, and it occurs at t = 0.

For the case where k_e is very small, as compared with k_s and h, (η >> 1, s >> 1), the roots of equation (10) are given by $p_n \sim n\pi$, and equations (7) and (8) transform to:

$$F = 1 + \frac{6\beta}{1-\beta} \sum \frac{1}{(n\pi)^2} \; e^{-\frac{\alpha \, (n\pi)^2 t}{R^2}} \qquad (21)$$

$$\Theta = \sum \frac{\sin n\pi x}{n\pi x} \; e^{-\frac{\alpha \, (n\pi)^2 t}{R^2}} \qquad (22)$$

The form of equation (21) is interesting. It shows that the uptake curve for a system controlled by heat transfer within the adsorbent mass has an equivalent mathematical form to that of the isothermal uptake by the Fickian diffusion model for mass transfer [26]. The isothermal model has mass diffusivity (D/R^2) instead of thermal diffusivity (α/R^2) in the exponential terms of equation (21). According to equation (21), uptake will be proportional to $\sqrt{t}$ at the early stages of the process which is usually accepted as evidence of intraparticle diffusion [27]. This study shows that such behavior may also be caused by heat transfer resistance inside the adsorbent mass. Equation (22) shows that the surface temperature of the adsorbent particle will remain at T_o at all t and the maximum temperature rise of the adsorbent is T^* at the center of the particle at t = 0. The magnitude of T^* depends on $(n_\infty - n_o)$, q, c and β, and can be very small in a differential test.

Experimental Verification

The proposed model was used to fit the uptake data for sorption of i-octane on 13X zeolite [8] and n-pentane on 5A zeolite [7]. These experiments were carried out using an assemblage of microsize crystals and satisfied the criteria for a differential test.

Figure 1 shows the uptake of i-octane on 13X zeolite at 403 K measured under a mean pressure level of 0.048 torr. The solid line (curve a) is the best fit of the data by equation (7) which shows that the data can be quantitatively described by the model. β for the system was evaluated using the equilibrium data [8]. A heat of adsorption of 19.5 kcal/mole was calculated for the system. c and ρ were taken to be 0.25 cal/gm/K and 0.8 g/cc, respectively. Curve b in Figure 1 shows the uptake curve for the same k_s as in curve a but with only the external film resistance controlling the heat transfer. Curve c shows the corresponding isothermal uptake. These curves clearly

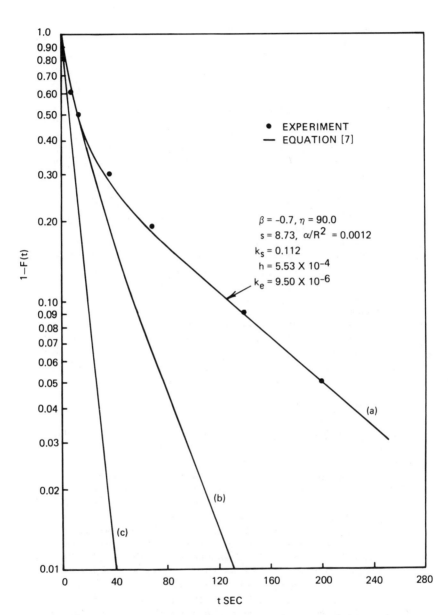

Figure 1. Sorption of i-octane on 13X at 403 K: (a) comparison of the experimental uptake with the best fit obtained by the present model; (b) uptake with no internal heat transfer resistance; and (c) uptake for an isothermal case.

demonstrate the importance and effects of various heat
transfer resistances on the shapes of the uptake curves.
They also show that ignoring any of these thermal resistances
in a model may yield a lower value of the mass transfer
coefficient.

Figure 2 shows the corresponding temperature profiles
at the center and the surface of the adsorbent mass calculated
by equation (8). The maximum temperature rises at the
center and the surface are, respectively, only 0.64°C and
0.23°C. Figure 2 also shows that the maximum temperature
rise occurs much earlier at the surface than at the center
and the surface temperature quickly approaches that of the
surrounding.

Figure 3 shows the uptake of n-pentane on 5A zeolite
[7] at 523 K (13 mg sample) measured under a mean presure
level of 24 torr. This system exhibits much faster uptake
than the previous one. The data can again be quantitatively
described by the model as shown by curve a in the figure.
The equilibrium data [28] was used to evaluate β for the
system. A heat of adsorption of 14.9 kcal/mole was estimated
for the system. Curves b and c, respectively, show the
corresponding uptakes for external film heat transfer control
and isothermal sorption using the same k_s as in curve a.

Figure 4 shows the calculated temperature profiles for
this system. The maximum temperature rises at the center
and the surface are, respectively, 2.3°C and 0.7°C and it
occurs earlier at the surface. The surface cools down to
near ambient temperature quicker than the center.

Effective Thermal Conductivity

No experimental data on the effective thermal conductivity
of an assemblage of micron size zeolite crystals under the
conditions of sorption tests used in the examples above
could be found in the literature. However, several methods
are available for the calculation of k_e for an assemblage of
particles with void fraction f. The thermal conductivities
of the solid phase (k_p) and the gas phase (k^o) in the voids
are needed [29]. We used the method developed by Maxwell.

k_g at low pressures can be significantly lower than the
gas thermal conductivity at atmospheric pressure (k^o) due to
the formation of Knudsen regime in the void space [30,31].
This is expected to happen when the mean free path (λ) of
the gas becomes comparable to the space between the particles
(ℓ_s) of the assemblage. Since the true size of the space
between the particles is not known, we assume it to be equal
to the diameter of the microparticle (d_p) as a first
approximation. The Knudsen effect has, however, been observed
at much higher pressures where $\lambda \ll d_p$ [32].

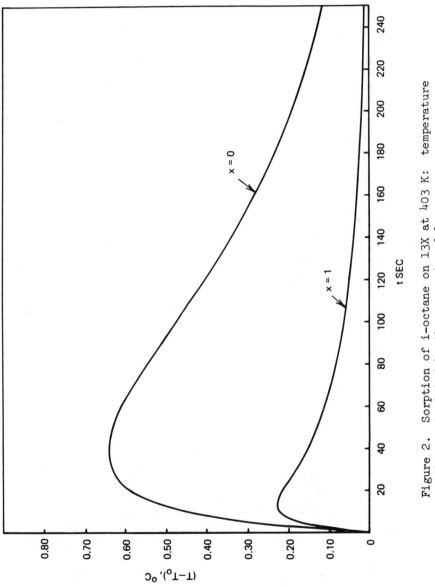

Figure 2. Sorption of i-octane on 13X at 403 K: temperature
profiles calculated by the present model.

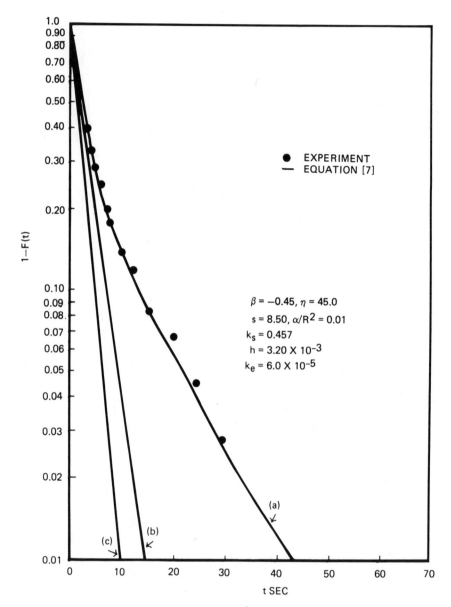

Figure 3. Sorption of n-pentane on 5Å at 523 K: (a) comparison of the experimental uptake with the best fit obtained by the present model; (b) uptake with no internal heat transfer resistance; and (c) uptake for an isothermal case.

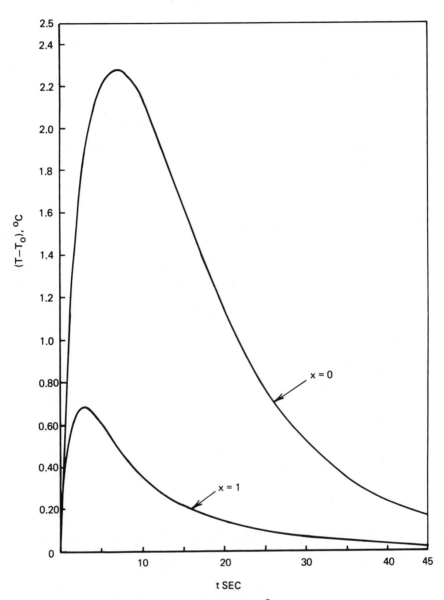

Figure 4. Sorption of n-pentane on 5Å at 523 K: temperature profiles calculated by the present model.

Following the procedure developed by Cha and McCoy
[31], the relationship between k_g in the Knudsen regime and
k_g^o for a polyatomic gas may be written as:

$$\frac{k_g}{k_g^o} = \left[1 + \frac{\lambda}{d_p}\ \frac{4 + 9\ (\gamma-1)}{A(\gamma+1)}\right]^{-1} \tag{23}$$

where γ is the ratio of the specific heats of the gas at
constant pressure and volume and A is a function of the
Knudsen thermal accomodation coefficients of the heat
exchanging surfaces. A depends on the nature of the gas
molecules and the surfaces [30]. k_g^o was estimated using the
method of Roy and Thodos and γ was obtained from the
published data [33].

The mean free paths of i-octane and n-pentane at the
conditions of the sorption experiments were calculated to be
750 and 2 microns, respectively. They are comparable with
the average crystal diameters of, respectively, 39 and 3.6
microns for the 13X and 5A zeolites used in the tests.
Using the above described procedure and $k_g = 4 \times 10^{-4}$
cal/sec/cm/K [34], f = 0.5 [8] and A = 1, we calculated k_e
of, respectively, 5×10^{-6} and 1.2×10^{-4} cal/sec/cm/K for
the i-octane-13X and n-pentane-5A systems. These numbers
are in good agreement at least in the order of magnitude
with the k_e values obtained from the analysis of sorption
data considering the uncertainties in the theories and the
values of the parameters (A, ℓ_s, f etc) used. It may be
concluded that k_e can be low for an assemblage of micro-
crystals at low pressures and high temperatures and internal
thermal resistance can be significant.

External Heat Transfer Coefficient

The external heat transfer coefficient is given by the
sum of the conductive (h_c) and radiative (h_R) contributions.
These depend on the shape and the size of the pan containing
the adsorbent, area of exposed adsorbents, volume ratio of
pan to adsorbent, emmisivities of the adsorbent and the pan
etc. We calculated an order of magnitude of h, for a sphere
with equivalent volume of the adsorbent mass by using
$h_c = k_g^o/R$ and $h_R = 4.33 \times 10^{-12}\ T^3$ [7]. The values for
i-octane-13X and n-pentane-5A systems were found to be,
respectively, 6.3×10^{-4} and 1.3×10^{-3} cal/cm^2/sec/K.
These values are in good agreement with those obtained from
the analysis of the sorption data by the LDF model in view
of the approximations made.

Parametric Study

Figure 5 shows the effect of η on the shape of the
uptake curve for a given s and β. Figure 6 shows the
corresponding dimensionless adsorbent temperature profiles
at x = 0. The abscisa in Figures 5 and 6 are dimensionless
time τ. Low η(low k_s, high k_e) corresponds to isothermal
sorption and very large η (high k_s, low k_e) represents
sorption under complete heat transfer control. The initial
slope of the uptake curve is a stong function of η but the
slope of the linear region at large τ is a weak function of
η indicating dominance of heat transfer in that region.
Figure 6 shows that the maximum temperature rise in the
adsorbent is higher for larger values of η and it occurs at
smaller values of τ. In real time, however, the maximum
occurs at longer times for lower k_e for a given η and the
adsorbent cools down slower.
Figure 7 shows the effects of the equilibrium parameter
β on the uptake curve for a given η and s. Fastest adsorption
occurs at β = 0 which corresponds to the isothermal case.
The rate of uptake slows down considerably as β decreases.
β affects the slopes of both the initial and the later
portions of the uptake curve. This shows that uptake rate
is very sensitive to the adsorbate loading and the heat of
sorption. For a Langmurian system, β may be written as:

$$\beta = - \frac{q\,\bar{n}^2\,\theta(1-\theta)}{c\,R_g\,T_o^2} \tag{24}$$

where $\bar{n}$ is the Langmuir saturation capacity and θ (= $n_\infty/\bar{n}$)
is the fractional coverage of the adsorbate in the test.
Equation (24) shows that β is a "U" shaped function of θ
having a minimum at θ = 0.50 and values approaching zero at
θ = 0 and 1. Consequently, the uptake will slow down as θ
increases from 0-0.5 and become faster after θ ≧ 0.50. This
would result in a "U" shaped dependence of diffusivity on
coverage estimated by an isothermal model. This behavior
has been experimentally observed [35].
Figure 8 shows the effect of β on the dimensionless
temperature profile at x =0. Since β is not an explicit
variable in equation (8), a single curve describes the
dimensionless temperature profile in terms of τ. The actual
temperature rise is lower as β decreases and the adsorbent
takes longer time to cool down.
Figure 9 illustrates the effect of the Biot number, s,
on the uptake curve for a given η and β. Figure 10 shows
the corresponding dimensionless temperature profiles at
x = 0. It may be seen from Figure 9 that s has small effect
on the uptake curve at low τ but it strongly affects the

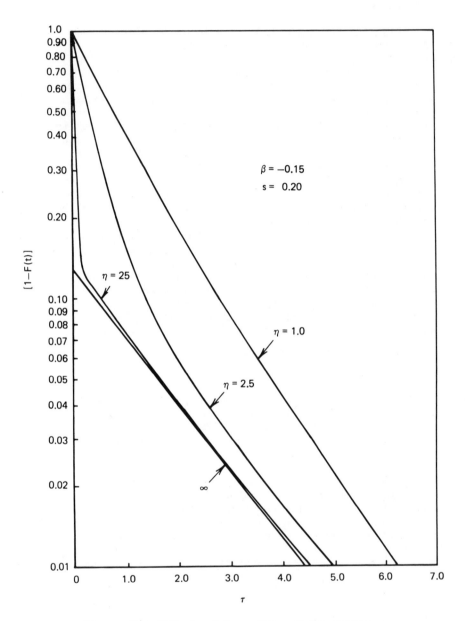

Figure 5. Effect of η on the uptake curves.

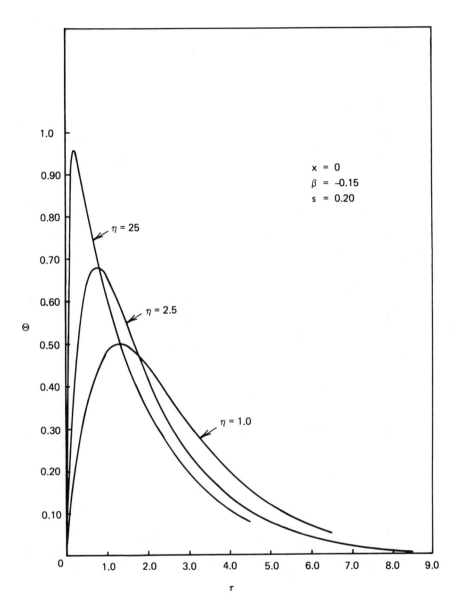

Figure 6.   Effect of η on dimensionless temperature profiles.

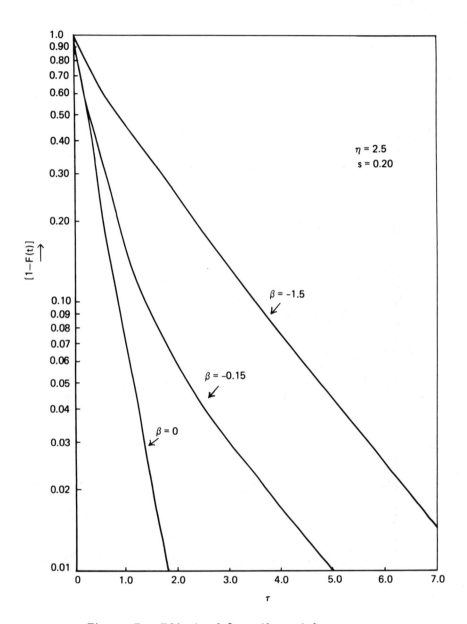

Figure 7.   Effect of β on the uptake curves.

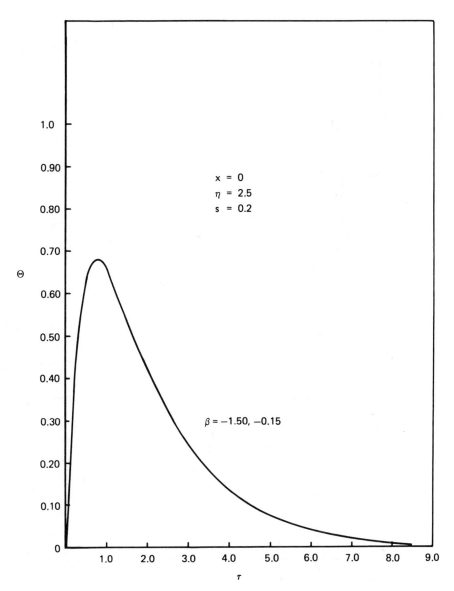

Figure 8. Effect of β on dimensionless temperature profiles.

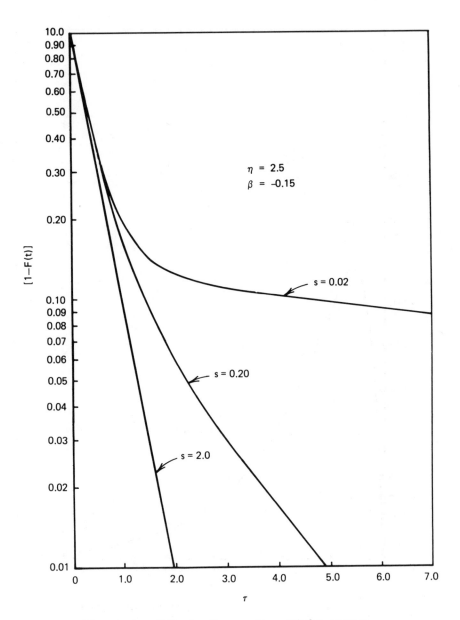

Figure 9.   Effect of s on the uptake curves.

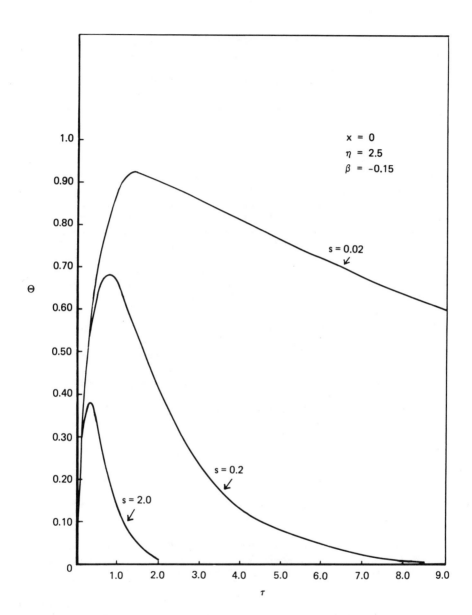

Figure 10.   Effect of s on dimensionless temperature profiles.

uptake at large τ. Low s significantly slows down uptake in τ domain. In real time, however, smaller k_e (large s) for a given k_s, β and h produces slower uptake. Figure 10 shows that the dimensionless temperature rise in the adsorbent is higher for smaller s (low h, high k_e). In real time the adsorbent cools down faster for larger k_e. The parametric study shows that the interactions between the thermodynamic and the mass and the heat transport properties of the adsorption system during the sorption process is quite complex.

Conclusion

The proposed model for non-isothermal sorption kinetics can quantitatively describe uptake$_0$data for adsorption of i-octane on 13X and n-pentane on 5A zeolites. The study indicates that the principal resistance to mass transfer for these systems may be confined at the surface of the zeolite crystals. It is also found that the internal thermal resistance of the assemblage of the micron size zeolite crystals used in the kinetic test is significant which produces a substantial thermal gradient within the assemblage and slows down the heat dissipation from it.

Thus, an apparently slower mass uptake is observed as compared with the isothermal uptake or the case when heat transfer is controlled by the external film resistance. Consequently, ignoring the internal thermal resistance may lead to erroneously low mass transfer coefficient.

It is recommended that careful experimental measurement of radial temperature-time profile in the adsorbent mass be carried out during mass uptake experiments for more accurate analysis of the process.

Literature Cited

1. Zolotarev, P. P.; Izv. Akad. Nauk. SSSR. Ser. Khim., 1970, 1421; 1970, 2831.
2. Zolotarev, P. P. and Radushkevich, L. V., Dokl. Akad. Nauk. SSSR, 1970, 195, 1361.
3. Zolotarev, P. P. and Kalinichev, A. I., Russ. J. Phys. Chem. (English Trans.), 1971, 45, 1610; 1972, 46, 654.
4. Brunovska, A., Hlavacek, V., Ilavsky, J., and Valtyni, J. Chem. Eng. Sci., 1978, 33, 1395; 1980, 35, 757; 1981, 36, 123.
5. Chihara, K., Suzuki, M., and Kawazoe, K., Chem. Eng. Sci., 1976, 31, 505.
6. Armstrong, A. A., and Stannet, V., Die Makromolekulare Chemie., 1966, 90, 145.

7. Ruthven, D. M., Lee, L. K., and Yucel, H., AIChE J., 1980, 26, 16.
8. Ruthven, D. M., Lee, L. K., AIChE J., 1981, 27, 654.
9. Kocirik, M., Smutek, M., Bezus, A., and Zikanova, A., Coll. Czech. Chem. Comm., 1980, 45, 3392.
10. Kärger, J., and Caro, J., J.C.S., Faraday I., 1977, 73, 1363.
11. Kärger, J., Caro, J., and Bülow, M., Izv. Akad. Nauk. SSSR. Ser. Khim., 1977, 2666.
12. Bülow, M., Struve, P., Finger, G., Redszus, C., Ehrhardt, K. Schirmer, W., and Kärger, J., J.C.S. Faraday I, 1980, 76, 597.
13. Lykow, A. W., Kolloidzeitschrift, 1935, 71, 333; 1936, 74, 179.
14. Wicke, E., ibid, 1939, 86, 167.
15. Timofeev, D. P., and Erashko, I. T., Russ. J. Phys. Chem. (English Trans.) 1971, 45, 359.
16. Ilavsky, J., Brunovska, A., and Hlavacek, V., Chem. Eng. Sci., 1980, 35, 2475.
17. Glueckauf, E., and Coates, J. J., J. Chem. Soc., 1947, 1315.
18. Thomas, W. J., and Lombardi, J. L., Trans. Inst. Chem., Engrs., 1971, 49, 240.
19. Sircar, S., and Kumar, R., I&EC Proc. Des. & Dev. In press.
20. Sircar, S., and Kumar, R., I&EC Proc. Des. & Dev. In press.
21. Kidnay, A. J., Hija, M. J., and Dickson, P. F., Adv. Cryogenic Eng., 1969, 15, 46.
22. Carter, J. W., and Hussain, H., Trans. Inst. Chem. Engrs., 1972, 50, 69.
23. Masamune, S., and Smith, J. M., AIChE J., 1965, 11, 34.
24. Yucel, H., and Ruthven, D. M., J. Colloid and Interface Sci., 1980, 74, 186.
25. King, G., and Cassie, A. B. D., Trans. Faraday Soc., 1940, 36, 445.
26. Crank, J., "The Mathematics of Diffusion", Oxford University Press, London, 1956; p. 86.
27. Barrer, R. M., "The Properties and Application of Zeolites", ed. R. P. Townsend, Chem. Soc. Special Publ. 33, 1979; p. 3.
28. Vavlitis, A. P., Ruthven, D. M., and Loughlin, K. F., J. Colloid and Interface Sci., 1981, 84, 526.
29. Krupiczka, R., Int. Chem. Eng., 1967, 7, 122.
30. Dushman, S., "Scientific Foundation of Vacuum Technique", John Wiley, New York, 1962; p. 45.
31. Cha, C. Y., and McCoy, B. J., J. Chem. Phys., 1972, 56, 3265.

32. Deissler, R. G., and Boegll, J. S., Trans., A.S.M.E.,
     1958, 1417.
33. Reid, R. C., Prausnitz, J. M., and Sherwood, T. K., "The
     Properties of Gases and Liquids", McGraw-Hill, New
     York, 1977, p. 481.
34. Lee, L. K., and Ruthven, D. M., J.C.S. Faraday I., 1979,
     79, 2406.
35. Ruthven, D. M., Separation and Purification Methods,
     1976, 5, 189.

RECEIVED January 4, 1983

Recovery and Purification of Light Gases by Pressure Swing Adsorption

HSING C. CHENG and FRANK B. HILL

Brookhaven National Laboratory, Upton, NY 11973

A cell model is presented for the description of
the separation of two-component gas mixtures by
pressure swing adsorption processes. Local equilibrium is assumed with linear, independent isotherms. The model is used to determine the light
gas enrichment and recovery performance of a
single-column recovery process and a two-column
recovery and purification process. The results
are discussed in general terms and with reference
to the separation of helium and methane.

Pressure swing adsorption (PSA) processes are widely
applied industrially for gas separations. Applications are
numerous and include hydrogen and helium recovery and purification, air drying, the production of oxygen from air, and the
separation of normal paraffins from isoparaffins.
   In spite of its widespread use, design information on PSA
processes has been rather sparse in the open literature. In
recent years, however, a beginning has been made in terms of the
development of mathematical models of PSA processes. This paper
is concerned with the further development of such models.
   Modeling efforts to date have been confined to two
processes: a single column recovery process and a two-column
recovery and purification process.
   The single-column process (Figure 1) is similar to that of
Jones et al. (1). This process is useful for bulk separations.
It produces a high pressure product enriched in light components. Local equilibrium models of this process have been
described by Turnock and Kadlec (2), Flores Fernandez and Kenney
(3), and Hill (4). Various approaches were used including
direct numerical solution of partial differential equations, use
of a cell model, and use of the method of characteristics.
Flores Fernandez and Kenney's work was reported to employ a cell
model but no details were given. Equilibrium models predict

0097–6156/83/0223–0195$06.00/0

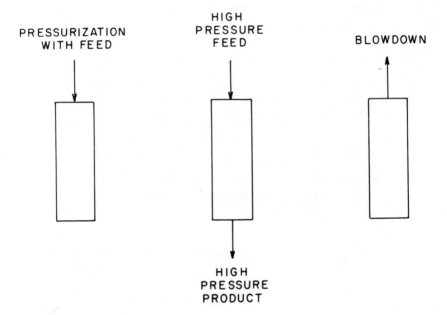

Figure 1.   Steps in single-column PSA process.

that this process is not capable of producing the light component in pure form.   Because of this fact the process is referred to here as a recovery process.

The two-column process (Figure 2) is the heatless adsorption process of Skarstrom (5).   This process can perform both recovery and purification of a light component.   Local equilibrium models of this process have been presented by Shendalman and Mitchell (6), Chan et al. (7), and Knaebel and Hill (8).   In each case the method of characteristics was used.   Models with finite mass transfer rates have been published by Kawazoe and Kawai (9), Mitchell and Shendalman (10), Chihara and Suzuki (11), and Richter et al. (12).   In these models the method of characteristics and direct integration of partial differential equations were employed.

Much of the modeling work to date has dealt with the separation of binary mixtures composed of a carrier and an impurity.   Such separation can be readily treated using the method of characteristics.   While in some situations (Knaebel and Hill, (8), for instance) the same method can be used for binary mixtures of arbitrary composition, the cell model is readily and generally useful for this situation.   It was this feature which prompted the use of the cell model in the present work.

In the present paper, a cell model is employed to simulate equilibrium isothermal PSA with a binary feed of arbitrary composition.   Linear isotherms are assumed.   The model equations derived are applied to the one- and two-column processes mentioned earlier to give a general description of their light component enrichment and recovery performance.   Also some discussion is given of the effects of various operating parameters on the separation of methane and helium.

Cell Model

A gas mixture consisting of two components is separated using one or more adsorption beds.   The beds are used in a cyclic process composed of steps involving column pressurization, depressurization (blowdown) and flow through the columns at constant pressure.   Analysis of such separation processes in terms of a cell model proceeds as follows.

Each adsorption bed is considered to consist of N well-mixed cells, each cell being of length ΔZ, as shown in Figure 3.   This figure depicts the notation used for process steps with flow from left to right.   The same notation can be used for steps with flow in the opposite direction if the cells are renumbered starting from the right, i.e., if

$$i' = N - i + 1 \qquad (1)$$

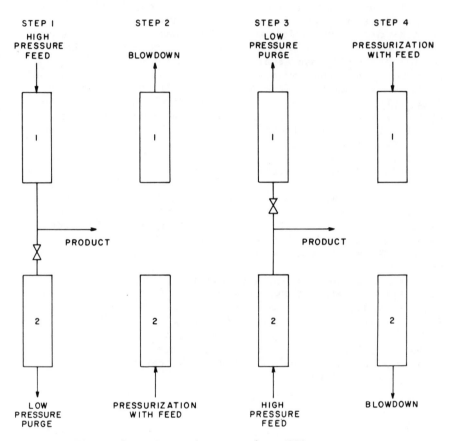

Figure 2.   Steps in two-column PSA process.

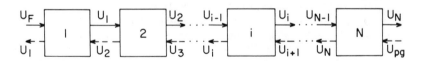

Figure 3.   Flow diagram for cell model.

Because of interchange between the gas and the adsorbent, the velocity and composition of the gas phase within the bed change with position and time during both constant pressure steps and steps with changing pressure. Pressure is assumed to vary with time during pressure changes but not spatially. Also, pressure drop due to flow is negligible. Finally, heat effects are neglected and local equilibrium with linear isotherms is assumed throughout the bed. These isotherms are represented by

$$n_\ell = k_\ell \frac{P}{RT} y \qquad (2)$$

$$n_h = k_h \frac{P}{RT} (1-y) \qquad (3)$$

where y is the mole fraction of the less strongly adsorbed or light component.

With the foregoing assumptions, the process steps of interest may be described starting with two material balances, a balance for the light component and a total balance. The balances are taken on the first and i-th cells and are written for a differential time dt.

The balances on the light component can be expressed as

$$\frac{d(Py_1)}{dt} = -\beta_\ell \frac{U_1 y_1 P - U_F y_F P_H}{\Delta Z} \qquad (4)$$

$$\frac{d(Py_i)}{dt} = -\beta_\ell P \frac{U_i y_i - U_{i-1} y_{i-1}}{\Delta Z} \qquad (5)$$

$$i = 2,3,\ldots,N$$

and the total balance yields

$$\frac{dP}{dt} = (1-\beta) \frac{d(Py_1)}{dt} - \beta_h \frac{U_1 P - U_F P_H}{\Delta Z} \qquad (6)$$

$$\frac{dP}{dt} = (1-\beta) \frac{d(Py_i)}{dt} - \beta_h P \frac{U_i - U_{i-1}}{\Delta Z} \qquad (7)$$

$$i = 2,3,\ldots,N$$

where

$$\beta_\ell = \frac{\varepsilon}{\varepsilon + (1-\varepsilon)\rho_s k_\ell} \qquad (8)$$

$$\beta_h = \frac{\varepsilon}{\varepsilon + (1-\varepsilon)\rho_s k_h} \qquad (9)$$

$$\beta = \frac{\beta_h}{\beta_\ell} \leq 1 \tag{10}$$

<u>Pressurization and Blowdown.</u> For these steps, Eqs. (4), (5), (6) and (7) are combined by eliminating $d(Py_1)/dt$ between them, respectively, obtaining

$$\frac{dP}{dt} = -\beta_\ell \frac{V_i P - V_F P_H}{\Delta Z} = -\beta_\ell P \frac{V_i - V_{i-1}}{\Delta Z} \tag{11}$$

where

$$V_F = [\beta + (1-\beta)y_F] U_F \tag{12}$$

and

$$V_i = [\beta + (1-\beta)y_i] U_i \tag{13}$$

$$i = 1,2,\ldots,N$$

For a pressurization step feed gas enters the first cell at velocity U_F and there is no flow from the N-th cell, i.e. $U_N = 0$. Therefore, when Eq. (11) is written for cell N and again for cell i, a difference equation in V_i is obtained whose solution is

$$V_i = \frac{(N-i)P_H}{NP} V_F \tag{14}$$

A dimensionless time is defined as

$$d\tau = \frac{\beta_\ell V_F P_H}{LP} dt \tag{15}$$

Using this definition and Eqs. (11), (12) and (15) in Eqs. (4) and (6), respectively, one obtains

$$\frac{dy_i}{d\tau} = -\left\{ \left[\frac{N-i}{\beta + (1-\beta)y_i} + 1 \right] y_i - \frac{N-i+1}{\beta + (1-\beta)y_{i-1}} y_{i-1} \right\} \tag{16}$$

$$i = 1,2,\ldots,N$$

with $y_0 = y_F$.

One can find the dimensionless time required to accomplish the full pressure change from P_L to P_H by substituting Eqs. (14) and (15) into Eq. (11) and integrating. The result is

$$\tau_P = \ln P_H/P_L \tag{17}$$

Thus the mole fraction profile after the pressure increase is calculated by integrating the N simultaneous differential equations represented by Eq. (16) from $\tau = 0$ to $\tau = \tau_P$.

For a blowdown step, gas leaves the bed through the N-th cell and there is no flow into the first cell, i.e., $U_F = 0$, and Eq. (11) written for cells N and i yields a difference equation in V_i whose solution is

$$V_i = \frac{i}{N} V_N \qquad (18)$$

The appropriate dimensionless time is

$$d\tau = \frac{\beta_\ell V_N}{L} \, dt \qquad (19)$$

Using this definition and Eqs. (11) and (18) in Eq. (6) one finds

$$\frac{dy_i}{d\tau} = - \left\{ \left[\frac{i}{\beta + (1-\beta)\, y_i} - 1 \right] y_i - \frac{i-1}{\beta + (1-\beta)y_{i-1}} \, y_{i-1} \right\} \qquad (20)$$

$$i = 1,2,\ldots,N$$

As before, one can show that the length of dimensionless time required for the pressure decrease during blowdown is that given by Eq. (17). The dimensionless time definition in this case is given by Eq. (19). Hence the mole fraction profile after the pressure decrease is found by integrating Eq. (20) from $\tau = 0$ to $\tau = \tau_p$.

Constant Pressure Step.   For treatment of this step Eqs. (4), (5), (6) and (7) can be combined, respectively, to yield

$$V_N = \ldots = V_i = V_{i-1} = \ldots = V_F \qquad (21)$$

With this result, Eqs. (6) and (7) can be written

$$\frac{dy_i}{d\tau} = \frac{\beta}{1-\beta} \, N \left[\frac{1}{\beta + (1-\beta)\, y_i} - \frac{1}{\beta + (1-\beta)y_{i-1}} \right] \qquad (22)$$

$$i = 1,2,\ldots,N$$

where dimensionless time is defined as

$$d\tau = \frac{\beta_\ell V_F}{L} \, dt \qquad (23)$$

Equations (22) are integrated over the interval $\tau = 0$ to $\tau = \tau_F$ where τ_F is any desired value.

Equations (16), (20) and (22) each are a set of N simultaneous first order differential equations which can be integrated to give the mole fraction profile within the adsorption bed as a function of the dimensionless time τ during the step

under consideration. The transient profiles for a given PSA
cycle are calculated by integrating these sets of equations
cyclically in the proper order.

Single Column Bulk Separation Process. A three-step single
column process was described by Hill (4) (Figure 1). The process is similar to the rapid PSA process of Jones et al. (1).
The cycle for this process consists of pressurization with feed
introduced through one end of a column followed by introduction
of additional feed through the same end at high pressure with
withdrawal of a product from the opposite end, and ending with
blowdown with the exhausted gas leaving via the feed end only.
Mole fraction profiles within the bed are calculated by integrating Eq. (16) from $\tau = 0$ to $\tau = \tau_p$ for the pressurization
step. An experimentally significant initial condition is that
of $y_i(0) = y_F$. The profile resulting from the integration of
Eq. (16) to time τ_p then becomes the initial condition for the
integration of Eq. (22) from $\tau = 0$ to $\tau = \tau_F$ for the feed
step. The profile for $\tau = \tau_F$ then becomes the initial condition for the integration of Eq. (20) from $\tau = 0$ to $\tau = \tau_p$ for
the blowdown step. Before performing the latter integration,
the cells and the corresponding mole fractions are renumbered in
accordance with Eq. (1) since the flow direction is reversed in
the blowdown step. Then the next cycle is started by renumbering the cells and mole fractions and using the mole fractions as
the initial condition for the next pressurization step. The
cyclic integration of Eqs. (16), (22) and (20) is then continued
until the cyclic steady state is reached.
    Two indices of process performance which are of interest
are the average steady state enrichment of the light component
in the product stream defined as

$$E = y_{PRD}/y_F \qquad (24)$$

and the steady state recovery of the light component in the
product stream

$$\rho = E\theta \qquad (25)$$

where the product cut, θ, is given by

$$\theta = \frac{N_{PRD}}{N_F + N_{PRS}} \qquad (26)$$

Since at the cylic steady state

$$N_F + N_{PRS} = N_{PRD} + N_{BD} \qquad (27)$$

Eq. (26) may be expressed as

$$\theta = \frac{N_{PRD}}{N_{PRD} + N_{BD}} \tag{28}$$

The product mole fraction, y_{PRD}, in Eq. (24) is the average steady state mole fraction leaving the adsorber during the feed step, i.e.,

$$y_{PRD} = \int_0^{\tau_F} \frac{y_N(\tau)}{\beta + (1-\beta)y_N(\tau)} \, d\tau \Big/ \int_0^{\tau_F} \frac{1}{\beta + (1-\beta)y_N(\tau)} \, d\tau \tag{29}$$

where $y_N(\tau)$ is the mole fraction transient for the N-th cell during the feed step in a steady state cycle and is obtained from the integration of Eq. (22).

It can be shown that N_{PRD}, the moles of gas taken off as product at cyclic steady state during the feed step is

$$N_{PRD} = \frac{\varepsilon A L}{\beta_{\ell}} \frac{P_H}{RT} \int_0^{\tau_F} \frac{d\tau}{\beta + (1-\beta)y_N(\tau)} \tag{30}$$

Also N_{BD}, the moles of gas taken from the column during the blowdown step at the cyclic steady state, is

$$N_{BD} = \frac{\varepsilon A L}{\beta_{\ell}} \frac{P_H}{RT} \int_0^{\tau_P} \frac{e^{-\tau} \, d\tau}{\beta + (1-\beta)y_N(\tau)} \tag{31}$$

where in this equation $y_N(\tau)$ is the mole fraction transient from the N-th cell during the blowdown step of a steady state cycle and is obtained by integration of Eq. (20).

Two-Column Process.   The heatless adsorption process (Figure 2) has two columns. Each column undergoes a four-step cycle. Three steps involve the same sequence as the single column process: pressurization with feed, feed at high pressure, and blowdown through the feed end. This is followed by a low pressure purge step with the purge being a fraction of the product from the other column and flowing from the product end to the feed end. For one column to provide purge for the other, the cycles in the two columns are 180° out of phase.

Modeling of this process proceeds in the same way as for the single-column process described above except for introduction of a purge step. The process is started with one column equilibrated with feed at high pressure and the other at low pressure. Eqs. (16), (20) and (22) are used as before with renumbering of cells and mole fractions at appropriate points in the cycle, but the purge step is inserted into the sequence at the appropriate point.

Three indices of performance for this process are of interest: E, ρ and H, the purge-to-total product ratio. The latter quantity takes the place of the purge-to-feed ratio in the heavy impurity-light carrier separation. For that separation it has been shown that the impurity is completely removed when the purge-to-feed ratio exceeds a critical value (7). When the heavy component content of the feed is above the impurity level, a ratio which is operationally more useful in determining whether a pure light gas can be obtained is the purge-to-total product ratio (8).

A total material balance around one column of the two-column process at cyclic steady state is

$$N_{PRS} + N_F + N_{PGF} = N_{PRD} + N_{BD} + N_{PG} \tag{32}$$

The extent of recovery of the light gas in the high pressure product is

$$\rho = \frac{(N_{PRD} - N_{PGF})}{N_{PRS} + N_F} E \tag{33}$$

It can be shown that H is given by

$$H = \frac{P_L}{P_H} \frac{\dfrac{\tau_{PGF}}{\beta + (1-\beta)y_{PGF}}}{\displaystyle\int_0^{\tau_F} \dfrac{d\tau}{\beta + (1-\beta)y_N(\tau)}} = \frac{P_L}{P_H} \frac{\tau_{PGF}}{\tau_F} \tag{34}$$

Results and Discussion

Eqns. (16), (20) and (22) were integrated numerically to obtain the separation performance of the one- and two-column processes. The GEAR package (13) was used for the integration after determining that it was faster than, say, Runge-Kutta methods. For all calculations $N = 50$ and $\varepsilon = 0.40$. Dimensionless parameters varied were β, P_H/P_L, y_F, and, for the two-column process, H. Combinations of the parameters of β and P_H/P_L were chosen to correspond to the methane-helium system on BPL carbon. Adsorption isotherm data for methane at 25°C (14) were represented by

$$n_h = \frac{5.69 \times 10^{-4} P_h}{1 + 8.11 \times 10^{-2} P_h} \tag{35}$$

where P_h is in atm and n_h is in gm mole/gm carbon. Helium was taken to be inert. The adsorption coefficient k_h used to calculate β was evaluated as the slope of the straight line connecting points on the isotherm, Eqn. (35), corresponding to the high and low pressures P_H and P_L. With values of β and P_H/P_L

related in this way, the calculated process performance can be described in terms of both its general features and its applicability to the methane-helium separation.

Single-Column Process. Typical relations showing enrichment and recovery of the light component in the product as a function of product cut are shown in Figure 4 with feed light component mole fraction as a parameter. Figures 4(a) and 4(b) are for high and low pressures of 60 and 30 psia, for which β = 0.083. For Figures 4(c) and 4(d) the high pressure is increased to 300 psia. Because of the leveling off of the methane isotherm, β increases to 0.156. For any feed mole fraction the enrichment is greatest at small cuts. This follows from the fact that pressurization leads to a spatial fractionation within the column with high mole fractions of the light component near the closed end. Small cuts then displace gas enriched in the light component into the product stream. Enrichment also increases as the mole fraction of the light component in the feed decreases. As a comparison of Figure 4(a) and 4(c) shows, increasing the high pressure produces an increase in enrichment. However, because of the leveling off of the methane isotherm the increase is not as large as that expected for a linear isotherm (4).

As seen in Figures 4(b) and 4(d), recovery increases rapidly with cut at small cuts and levels off at large cuts. Small light component mole fractions in the feed and large pressure ratios favor high recoveries.

An example of the effect of pressure on the performance of the single column process for the methane-helium separation is shown in Table 1. Using Figure 4(b) and 4(d), the cuts corresponding to 80 percent recovery of helium were determined as a function of y_F and P_H/P_L. Then the enrichments and in turn the product mole fractions corresponding to these cuts were found from Figures 4(a) and 4(c). Except for the leanest feed mole fraction, y_F = 0.005, the use of the higher top pressure leads to a significantly higher product mole fraction.

Two-Column Process. Typical enrichment-cut and recovery-cut relations for this process are shown in Figure 5. For this figure P_H = 60 psia, P_L = 30 psia, and β = 0.083. For Figures 5(a) and 5(b), y_F = 0.005 and for Figures 5(c) and 5(d), y_F = 0.1. The parameter in Figure 5 is H, the purge-to-product ratio.

The maximum enrichment possible for the 0.5 percent helium feed is 200. For 10 percent helium it is 10. Pure helium is obtained with the 10 percent feed with H = 0.8 at cuts smaller than 0.018. At smaller values of H, pure helium cannot be produced. Very large enrichments are obtained at small cuts for 0.5 percent helium but pure helium is not obtained. This finding is consistent with the results of Knaebel and Hill (8) who

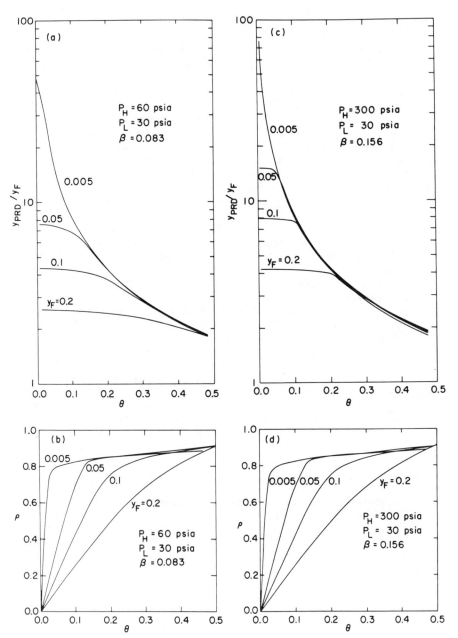

Figure 4. Light gas enrichment and recovery in product as a function of product cut in single-column PSA process.

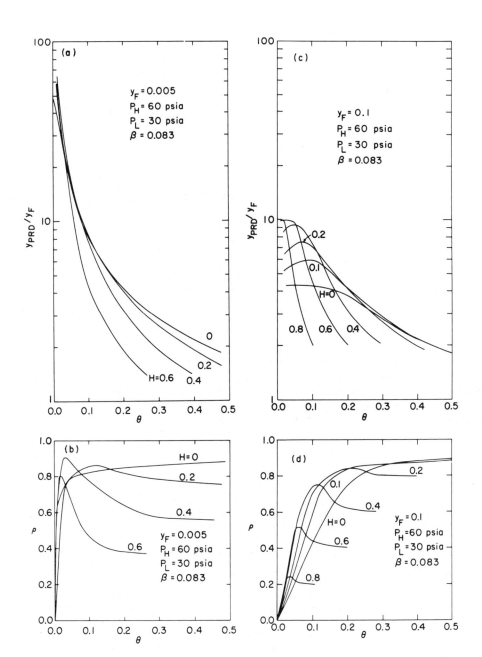

Figure 5. Light gas enrichment and recovery in product as a function of product cut in two-column PSA process.

Table I

Effect of Pressure on Single-Column
Process for Helium-Methane Separation

$\rho = 0.80$

y_F	P_H/P_L	θ	E	y_{PRD}
0.005	2	0.057	14	0.070
	10	0.063	13	0.065
0.05	2	0.128	8.2	0.41
	10	0.063	13	0.65
0.1	2	0.238	3.4	0.34
	10	0.107	7.5	0.75
0.2	2	0.364	2.13	0.43
	10	0.203	3.95	0.79

found that a critical pressure ratio (as well as a critical H) had to be exceeded to obtain pure light gas and that this ratio was higher the smaller the light gas content of the feed. The critical ratios for the conditions of Figure 5 cannot be calculated from the results of Knaebel and Hill since their results were obtained for the "shortest clean column" condition. But evidently a pressure ratio of 2.0 is high enough for perfect cleanup for a 10 percent helium feed but not high enough for a 0.5 percent feed.

The curves for H = 0 represent the performance of the single-column process. It is apparent that the single-column process is superior at large product cuts in terms of both enrichment and recovery. For small cuts, the two-column process is superior.

The effect of feed composition on enrichment at maximum recovery is shown in Table 2. Both high enrichment and good recovery are obtainable with the leaner feed composition, a result which is counter to intuition. The explanation lies in the fact that light gas losses in the purge and blowdown gas decrease as the feed becomes leaner in the light gas.

Table II

Enrichment at Maximum Recovery
Two-Column Process

$\beta = 0.083$ $P_H/P_L = 2$

H	ρ_{max}	$\theta_{\rho=\rho_{max}}$	$E_{\rho=\rho_{max}}$
	$y_F = 0.005$		
0.0	1.0	1.0	1.0
0.2	0.863	0.122	7.0
0.4	0.905	0.032	24.3
0.6	0.800	0.017	50.0
	$y_F = 0.1$		
0.0	1.0	1.0	1.0
0.2	0.835	0.2	4.28
0.4	0.750	0.115	6.5
0.6	0.522	0.062	8.4
0.8	0.244	0.037	6.6

Conclusion

The equations for a local equilibrium cell model of pressure swing adsorption processes with linear isotherms have been derived. These equations may be used to describe any PSA cycle composed of pressurization and blowdown steps and steps with flow at constant pressure. The use of the equations was illustrated by obtaining solutions for a single-column recovery process and a two-column recovery and purification process. The single-column process was superior in enrichment and recovery of the light component at large product cuts. The two-column process was superior at small cuts.

Acknowledgments

This work was supported by the Division of Chemical Sciences, U.S. Department of Energy, Washington, D.C., under Contract No. DE-AC02-76CH00016.

Notation

E enrichment of ligh gas in product stream
H molar purge-to-total product ratio
i cell number
L column length
n concentration of sorbate in solid phase
N number of moles entering or leaving adsorber per cycle or total number of cells
P gas pressure
R gas constant
t time
T temperature
U interstitial gas velocity in bed
V concentration velocity defined in Eq. (13)
y gas phase mole fraction of light gas
ΔZ cell length

Greek symbols

β separation factor β_h/β_ℓ
β_i equilibrium ratio of gas capacity of component i to total capacity of gas and solid phases for component i
ε void fraction
ρ recovery of light gas
ρ_s density of solid particles
θ product cut, moles of product per mole of feed
τ dimensionless time

Subscripts

BD refers to blowdown
F refers to feed
h refers to heavy gas
H refers to high pressure
ℓ refers to light gas
L refers to low pressure
PG refers to purge
PRD refers to high pressure product
PRS refers to pressurization

Literature Cited

1. Jones, R. L.; Keller, II, G. E.; Wells, R. C., U.S. Patent 4,194,892, March 25, 1980.
2. Turnock, P.; Kadlec, R. H. A.I.Ch.E. J. 1971, 17, 335.
3. Flores Fernandez, G.; Kenney, C. N. paper presented at San Francisco A.I.Ch.E. Meeting, November, 1979.
4. Hill, F. B. Chem. Eng. Commun. 1980, 7, 37.
5. Skarstorm, C. W. Ann. N.Y. Acad. Sci. 1959, 72, 751.
6. Shendalman, L. H.; Mitchell, J. E. Chem. Eng. Sci. 1972, 27, 1449.
7. Chan. Y. N. I.; Hill, F. B.; Wong, Y. W. Chem. Eng. Sci. 1981, 36, 243.
8. Knaebel, K. S.; Hill, F. B. paper presented at Washington A.S.M.E. Meeting, November, 1981.
9. Kawazoe, K.; Kawai, T. Kagaku Kogaku 1973, 3, 228.
10. Mitchell, J. E.; Shandalman, L. H. A.I.Ch.E. Symp. Ser. 1973, 39, 134, 23.
11. Chihara, K.; Suzuki, M. paper presented at 2nd World Congress of Chem. Eng., Montreal, Canada, October, 1981.
12. Richter, E.; Knoblauch, K.; Jüntger, H. paper presented at the 7th Int. Congr. of Chem. Eng. Chem. Equipment Des. and Automation, Chisa '81, Praha, Czechoslovakia, Aug. 31-Sept. 4, 1981.
13. Hindmarsh, A. C. "GEAR: Ordinary Differential Equation System Solver"; Lawrence Livermore Laboratory, Report UCID-30001, Revision 3, December, 1974.
14. Grant, K. J.; Manes, M. I.E.C. Fundamentals 1966, 5, 490.

RECEIVED December 28, 1982

Methane/Nitrogen Gas Separation over the Zeolite Clinoptilolite by the Selective Adsorption of Nitrogen

T. C. FRANKIEWICZ and R. G. DONNELLY

Occidental Research Corporation, PO Box 19601, Irvine, CA 92713

N_2 diffuses into the structural pores of clinoptilolite 10^3 to 10^4 times faster than does CH_4. Thus internal surfaces are kinetically selective for N_2 adsorption. Some clino samples are more effective at N_2/CH_4 separation than others and this property was correlated with the zeolite surface cation population. An incompletely exchanged clino containing doubly charged cations appears to be the most selective for N_2. Using a computer-controlled pressure swing adsorption apparatus, several process variables were studied in multiple cycle experiments. These included feed composition and rates, and adsorber temperature, pressure and regeneration conditions. N_2 diffusive flux reverses after about 60 seconds, but CH_4 adsorption continues. This causes a decay in the observed N_2/CH_4 separation. Therefore, optimum process conditions include rapid adsorber pressurization and short adsorption/desorption/regeneration cycles.

Nitrogen is weakly adsorbed on the surfaces of numerous zeolites. The surface attraction arises as a result of the nuclear induced quadrupole moment possessed by N_2. The effect is weak and heats of adsorption for N_2 are low. Methane is a weakly polarizable molecule which, as a result of this polarizability, is also weakly adsorbed on zeolites. In general, H_{ADS} for CH_4 is equal to or slightly greater than H_{ADS} for N_2 and as a result an adsorption separation of these gases either does not occur or happens by a weakly selective CH_4 adsorption. In the 1950s, Habgood (1) discovered that he could selectively adsorb N_2 from a CH_4/N_2 mixture by cooling a 4A molecular sieve to between 0° and -78°C. However, his system was limited by slow kinetics, a weak selectivity, and the need to "thermally" regenerate the sorbent.

0097–6156/83/0223–0213$06.25/0

Recently, we found that the naturally occurring zeolite clinoptilolite could selectively adsorb N_2 in the presence of CH_4 near ambient temperature. This paper will show the selectivity to be kinetic rather than thermodynamic and that only clinoptilolite (clino) with the appropriate surface cation population will effect the separation. In addition, we review our results from several series of pressure swing CH_4/N_2 separation experiments and show how to design a process to use this effect. An economic evaluation suggested the process would be economically viable for small plants $(2 \times 10^6 \ ft^3/day)$.

Characterization of Clinoptilolite

Clinoptilolite is an abundant, naturally occurring zeolite whose structure has recently been determined (2,3). The clino structure is nearly identical to that of heulandite. A combination of lower Al^{+++} in the framework and an extra stable cation position appears to render clino thermally stable relative to heulandite.

Clino has a layer structure with the AC or (010) plane being impervious to gaseous diffusion of any kind. Parallel to the C-Axis (001) are alternating channels defined by 8- and 10-membered framework oxygen atom rings. The 8-member ring channel dimensions are about 4.6 x 4.0 Å while the larger channels are 7.0 x 4.3 Å. Parallel to the a-axis (100) are uniform 5.4 x 3.9 Å channels defined by 8-membered framework oxygen atom rings (4).

Early work on the catalytic behavior of clinoptilolite found that clinos from different geographic locations generally possessed widely differing properties (5). The most likely explanation is that naturally occurring clinos possess varying degrees of cation deficiency and that this deficiency variation results in a significantly altered pore accessibility. This property also leads to the unique CH_4/N_2 molecular sieving capability of this mineral. Table 1 shows how BET surface areas and CH_4/N_2 separations ability correlate with the degree of ion exchange in 4 clino samples.

The zeolites designated ZBS-14 and ZBS-15 are clinos from two separate deposits owned by an Occidental Petroleum subsidiary, Occidental Minerals, Inc. H-ZBS-15 is a decationized version of its parent. Ash Meadows is an ARCO owned deposit and represents a completely ion exchanged clino sample.

Heat of Adsorption Measurements

Using a gas chromatographic technique (6-11), apparent heats of adsorption

$$H_{ADS}(\text{apparent}) = H_{ADS}(\text{actual}) + E_a(\text{diffusion}) \qquad (1)$$

were measured for Ar, N_2, and CH_4 on the samples listed in Table I
and on several decationized and partially ion exchanged versions
of the parent minerals. These results are listed in Table II.
Note that for the most open clino, decationized ZBS-15, the heats
of adsorption for CH_4 and N_2 are the same. As the zeolite is
progressively re-cationized, the accessibility of CH_4 to the
internal crystal surfaces progressively decreases. This is
reflected in a lower determination for H_{ADS} (12). Finally, in
the completely cation exchanged ash meadows sample, even H_{ADS} for
N_2 begins to show a decrease. The conclusion is that given equal
access to internal surfaces, CH_4 and N_2 will be equally adsorbed.
However, molecular sieving can still effect a separation. Based
upon an analysis of the H_{ADS} measurements and the g.c. peak
shapes obtained, we can make a semi-quantitative assessment of the
structural pore accessibility for several gases into the zeolites
listed in Table II. This is shown in Figure 1. The shaded band
represents the desired pore access cut-off for maximum CH_4/N_2
separation.

Capacity and Kinetics for CH_4/N_2 Adsorption

Given the ability of the clino ZBS-15 to effect CH_4/N_2 separation chromatographically, we measured a high pressure adsorption
isotherm for N_2 and diffusion kinetics for CH_4 and N_2. A complete
isotherm for CH_4 was not taken due to slow kinetics, but at low
pressures ($P_{CH_4} <$ 50 psia), CH_4 and N_2 capacities appeared to be
similar. The apparatus used for both measurements is shown sche-
matically in Figure 2. Gas adsorption capacities and rates are
determined by following gas pressure vs. time. A Van der Waals
equation of state is used to calculate total number of moles of
gas in the ballast prior to adsorption and in the ballast and
adsorber after adsorption is complete. The difference in these
numbers is the adsorbed gas and the rate at which gas is adsorbed
is directly related to the diffusion coefficient by equation (2).
The apparatus is monitored and controlled by a Digital Equipment
Company laboratory computer (MINC). During the determination of
diffusion coefficients in a mixed gas system, a sample line is
installed and bed samples analyzed on a quadrupole mass spectrometer. The mass spec is also monitored by the MINC with data
acquired at 16 msec. intervals.

The adsorption isotherm for N_2 onto ZBS-15 at 20°C is shown
in Figure 3.

The diffusion coefficient for a gas into a solid may be
determined by measuring P vs t and making the plots exemplified by
Figures 4(a) and 4(b). In Figure 4, the relative amount of gas
adsorbed in the experiment, R from equation (2), is plotted vs.

Table I

Both Surface Areas and the Ability of Clinoptilolite
to Effect N_2/CH_4 Separation Are Correlated
With How Completely the Sample Is Ion Exchanged.

Clino Sample	Total Ion Charge / Al Content	Surface* Area	Ability to Separate N_2/CH_4
H-ZBS-15	.12	71.9 m^2/g	None
ZBS-15	.38	52	Good
ZBS-14	.88	36	Modest
Ash Meadows	1.0	12	None

* N_2 BET measurements

Table II

Apparent Heats of Adsorption for Several
Gases Over Various Clinoptilolites
(Kcal/Mole)

	Apparent Heat of Adsorption		
Molecule / Kinetic Diameter	Ar 3.40Å	N_2 3.64Å	CH_4 3.8Å
Zeolite			
ZBS-14	0	5.0	0
K-ZBS-14*	4.2	5.1	5.1
ZBS-15	(-)	5.2	0
H-ZBS-15	4.2	5.3	5.1
Ca-ZBS-15*	3.8	5.9	1.2
K-ZBS-15*	3.7	5.3	0.8
Ash Meadows	(-)	3.8	0

*Partially Cation Exchanged
(-) not measured

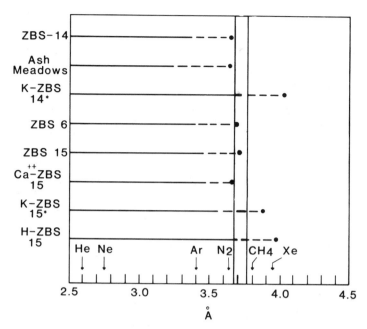

Easy Pore Access (———)
Hindered Pore Access (– – – –)
Approx. Size Cut-Off for Pore Access (•)

* Partially Cation Exchanged

Figure 1. A semi-quantitative assessment of structural pore accessibility for several clino samples to a range of molecular kinetic diameters.

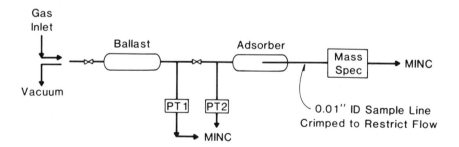

PT = Pressure Transducer

Figure 2. A schematic drawing of the apparatus used for the measurement of CH_4 and N_2 diffusion coefficients.

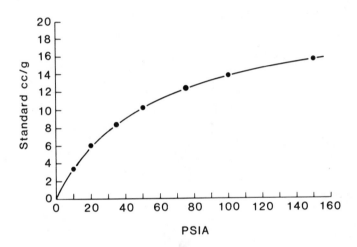

Figure 3. The adsorption isotherm (21 $^{\circ}$C) for N_2 on the clinoptilo-lite designated ZBS-15.

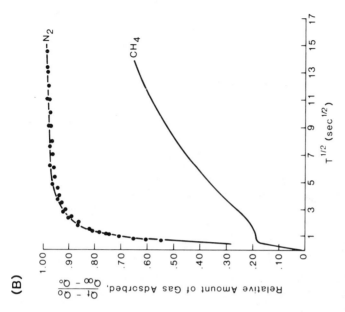

(B)

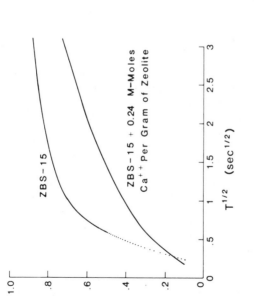

(A)

Figure 4. Representative curves are shown for the measurement of diffusion coefficients. (A) Curves are for N_2 adsorption onto the clino ZBS-15 and calcium impregnated ZBS-15. (B) Curves are for CH_4 and N_2 adsorption onto ZBS-15.

the square root of time. At sufficiently short times, the following equation is valid (13):

$$R = \frac{Q_t - Q_0}{Q_\infty - Q_0} = 2\frac{A}{V} \cdot \frac{1+K}{K} \cdot \sqrt{\frac{Dt}{\pi}} \quad (2)$$

$$K = \frac{(Q_0)_g - (Q_\infty - Q_0)}{Q_\infty} \quad (3)$$

where R   = relative amount of gas adsorbed during run
       Q_0 = amt. of gas on the zeolite at t = 0
       Q_t = amt. of gas adsorbed at any time t
       Q_∞ = final amt. of gas adsorbed
       $(Q_0)_g$ = amt. of gas initially available for adsorption
       D   = diffusion coefficient in cm^2/sec
       A/V = surface to volume ratio of the particle which is controlling diffusion

     Representative results from diffusion measurements are illustrated in Figures 4(a) and 4(b) and tabulated in Table III. The "knee" in the CH_4 data at 0.2 relative adsorption indicates that about 20% of the total adsorption surface is rapidly accessed and is thus non-selective in CH_4/N_2 adsorption. Once this "external" surface region is saturated, adsorption occurs only in the crystallite structural pores, which are kinetically selective for N_2. At very long times, adsorption approaches thermodynamic equilibrium, and there will be a tendency for the extra N_2 residing in structural pores to be replaced with CH_4. Therefore, maximum separation at short times is predicted (t<9-25 sec. in the case of ZBS-15), and separation will be higher in a process if subsequent gas-solid separation steps (e.g., depressurization) follow quickly. Significantly, measured diffusion coefficients were found to be independent of system pressure over a range of final pressures from 1 to 200 psia. Thus, the results presented here are likely to be applicable to real operating systems. Note that the addition of Ca^{++} to the ZBS-15 has a significant impact on the diffusion of both N_2 and CH_4. The impact on N_2 is greater, however, and the ratio of the two diffusion coefficients drops somewhat. How this change affects pressure swing separation of the two gases will be discussed below.

Diffusion in Mixed Gas Systems

     Since methane diffusion is much slower than nitrogen diffusion into clinoptilolite, it is reasonable to suspect that the presence of CH_4 on the zeolite surface will effectively slow the uptake of N_2. If true, then it would be much more appropriate to measure D_{N_2} in the presence of CH_4 rather than in the pure gas system. To accomplish this, the apparatus of Figure 2 was modified by the addition of a 0.01" i.d. sample line which terminated

Table III

Summary Tabulation of Diffusion Coefficient
Measurements on ZBS-15 and Ca^{++} Impregnated
ZBS-15, A/V = 1.67×10^4

Particle Size	Gas	Final System Pressure	D_{N_2}	D_{CH_4}	D_{N_2}/D_{CH_4}
As received ZBS-15					
60 x 80 mesh	N_2	27 psia	2.3×10^{-10} cm^2/sec		
"	N_2	74	1.9×10^{-10}		
20 x 50	N_2	7	4×10^{-10}		
6 x 10	CH_4	11		1.2×10^{-14}	2.0×10^4
Ca^{++} impregnated ZBS-15					
20 x 50 mesh	N_2	12 psia	1.8×10^{-11} cm^2/sec		
20 x 50	N_2	172	1.7×10^{-11}		
20 x 50	CH_4	11		5.8×10^{-15}	3.1×10^3

in the center of the zeolite adsorber. Adsorber gas composition
was monitored by the MINC computer as analyzed every 16 msec. by
the quadrupole mass spectrometer. System pressures were measured
with Validyne pressure transducers and also monitored by the MINC.
From this data, % N_2, % CH_4, and partial pressures of each gas can
be calculated vs. time.

Figure 5 illustrates the mixed gas diffusion data for a pair
of runs using 60% CH_4/40% N_2 as the feed gas. Note that N_2
adsorption is a maximum at about 10 seconds for ZBS-15 while CH_4
adsorption continues even after 15 minutes. It is important to
realize that since N_2 diffusion reverses after 10 seconds, Q in
equation (2) may be low and result in an artificially high value
being calculated for D_{N_2}. With this caveat, $D_{N_2} \leq 5 \times 10^{-11}$
cm^2/sec for this mixed gas system over ZBS-15 (as received).
Thus, the presence of methane in this gas mixture has reduced
D_{N_2}/D_{CH_4} from 2.0 x 10^4 (see Table III) to 3.3 x 10^3. Table IV
presents data for other gas mixtures with ZBS-15 and CA^{++} treated
ZBS-15. Methane appears to have a lesser effect on N_2 diffusion
in the Ca^{++} treated ZBS-15, but the calculational caveat given
above applies here as well.

The real impact of the difference in mixed gas diffusion
rates for the two zeolites becomes apparent in Figure 6. Here
Separation Factors vs Time are plotted for the two adsorbents.
Note that for ZBS-15, maximum separation is observed at 20 seconds
and has decayed to less than 1/2 of its maximum in the 90-180 sec.
time frame which characterized most of the pressure swing experiments to be discussed below. The separation over Ca^{++} treated
ZBS-15 also decays but is still much improved over the as-received
material even after 3 minutes.

The implications of Figure 6 for process design are significant. Using ZBS-15, a mean space time for an adsorber of about 20
seconds is optimum. For Ca^{++} treated ZBS-15, this number can be
increased to 60-120 seconds with no loss of separation relative to
the parent zeolite. Due to very low initial pressure in the
regenerated adsorber, the high pressure feed gas will expand and
move very rapidly through the adsorber bed. As a result of this
very short residence time (near t=o in Figure 6), an initially
poor separation CH_4/N_2 is predicted. Thus, initial adsorber
effluent will be high in N_2 and may have to be discarded or
recycled.

The Pressure Swing Separation of CH_4 and N_2

In order to investigate the process aspects of the pressure
swing separation of CH_4 and N_2 the apparatus depicted in Figure 7
was constructed. Valves V1 through V5 are fitted with electric
switching mechanisms that are interfaced to the MINC computer.
The two pressure transducers, three mass flow meters, and column

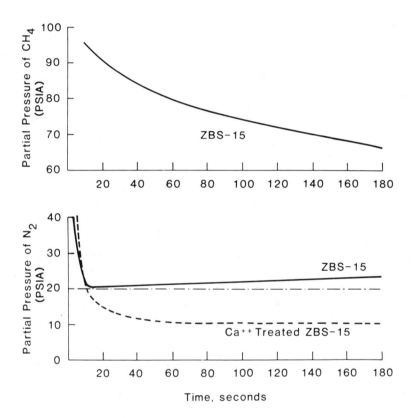

Figure 5. Partial pressure vs. time curves show how the addition of Ca^{++} to ZBS-15 improves its ability to retain N_2.

Table IV

Diffusion Coefficients Were Measured <u>Vs.</u> Gas Composition for (1) ZBS-15 and (2) ZBS-15 + .24 m moles/g of Ca^{++}. The Presence of CH_4 Slows N_2 Diffusion.

Zeolite	Feed Gas	D_{N_2}	D_{N_2}/D_{CH_4} *
(1) ZBS-15	N_2	23×10^{-11} cm^2/sec	2.0×10^4
	60% CH_4	5×10^{-11}	
	80% CH_4	4×10^{-11}	3.3×10^3
	90% CH_4	4×10^{-11}	
(2) ZBS-15 +	N_2	1.8×10^{-11} cm^2/sec	3.1×10^3
.24 m moles Ca^{++}/g	80% CH_4	1.2×10^{-11}	3.0×10^3
	90% CH_4	1.3×10^{-11}	

---

* D_{CH_4} is assumed to be unchanged from the pure gas measurement

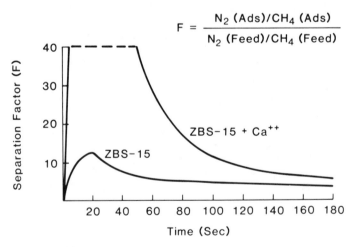

$$F = \frac{N_2 \,(Ads)/CH_4 \,(Ads)}{N_2 \,(Feed)/CH_4 \,(Feed)}$$

Figure 6. Ca^{++} impregnated ZBS-15 shows an enhanced ability to separate CH_4 and N_2.

V = Electrically Operated Valve
PT = Pressure Transducer
FM = Mass Flow Meter
BPR = Back Pressure Regulator
C = Adsorption Column
F = In-line Gas Filter

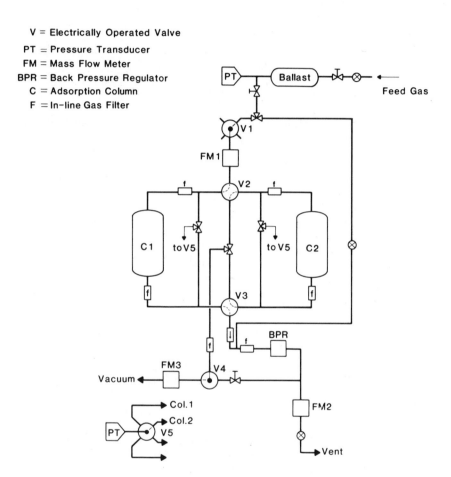

Figure 7. A schematic drawing of the computer-controlled pressure swing apparatus.

temperature are also monitored by the MINC. All gas effluent from the apparatus through vent or vacuum is sampled through a molecular leak valve for analysis by the quadrupole mass spectrometer. The mass spec in turn reports these compositions to the MINC. Once a run is initiated, the apparatus is under complete control of and monitored by the MINC.

The back pressure regulator (BPR) is set to give the desired pressure in the adsorption column (C1 or C2) being used. Although it is possible with this apparatus to involve both columns alternately in a separations experiment, this was not done during our runs so that all vent and vacuum regeneration gases could be analyzed for every cycle in the run. During a pressure swing run, gas enters the adsorption column, C1 or C2, through valve V2. Valve V3 is set to allow gas to exit the selected adsorber through the back pressure regular (BPR) as soon as the desired operating pressure is achieved. For depressurization, V3 is switched to allow the column exit gas to bypass the BPR and be vented through valve V4. V4 is a vacuum/vent selector and V5 allows a single pressure transducer to monitor system pressures at up to four selected points.

A sample data plot is shown in Figure 8. Feed gas is 40% N_2 and 60% CH_4. A mass balance is calculated for CH_4 and N_2 for each step in the process as well as for the complete adsorption cycle. The mass balance includes a Van der Waals calculation of the amount of gas stored in column voids and the volume of gas adsorbed on the zeolite. Typically, the independently calculated mass balances for CH_4 and N_2 were 100   3%.

Table V lists the independent variables and their respective ranges which were included in this study. Dependent variables can be defined in numerous ways depending upon the objectives of the particular experimental series being conducted. Typically, a pressure swing adsorption cycle program is input to the apparatus's computer control program, e.g., per Figure 8, and the cycle repeated under computer control until the experimental objective is achieved. For this work, 6-24 cycles were normally used with total run times up to 8 hours.

Figures 9 and 10 illustrate changes in two dependent variables: dynamic N_2 adsorption capacities and CH_4/N_2 separation factors. Independent variables are column temperature, operating pressure, and time allowed for vacuum regeneration. This experimental series used a constant feed rate of 6.0 l/min over a time of 1.00 min into a 1" dia. x 24" long adsorber filled with 180g of zeolite. Column depressurization took place for 1.00 min. and this was followed by a variable length vacuum regeneration.

Adsorbent particle sizes, although not a design variable, did affect some results for hydrodynamic reasons. As can be seen from the centerpoints, a slight deterioration in column performance resulted with the switch from a 60 x 80 to a 6 x 10 mesh adsorbent. Although this effect was real and reproducible, the effects of the three design variables were generally more significant.

(1) Depressurization Begins

(2) R Vacuum Regeneration Begins

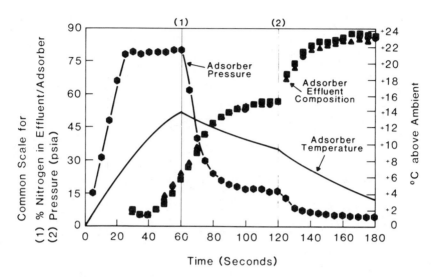

Figure 8.   Adsorber parameters are plotted vs. time for a pressure swing run.   Note the common scale for $\%N_2$ and adsorber pressure.

Table V

Operational Variables and Variable Ranges Studied for the Pressure Swing Separation of CH_4 and N_2

Gas Feed Rate	16.7 - 50.0 scc/g. of zeolite-min.
Feed gas composition	10 - 40% N_2
Adsorber Temperature	0° - 55°C
Operating Pressure	75 - 300 psia
Adsorber Granule Size	.18 - 3.5 mm
Column Regeneration	0 - 20 min. vacuum or CH_4 purge

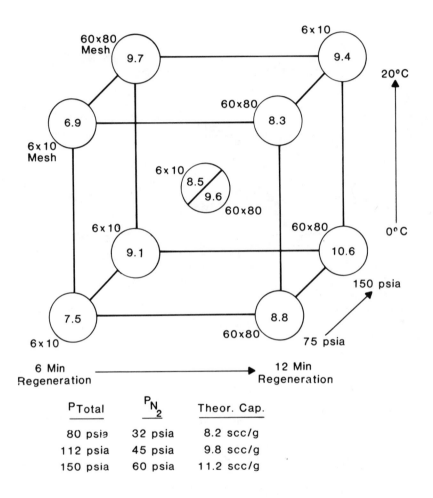

Figure 9. Average SCC of N_2 adsorbed per gram of the clino ZBS-15 for a 3 variable factorial design experimental series. Feed was 33.3 SCC/g of zeolite per cycle of 40%N_2/60%CH_4 gas.

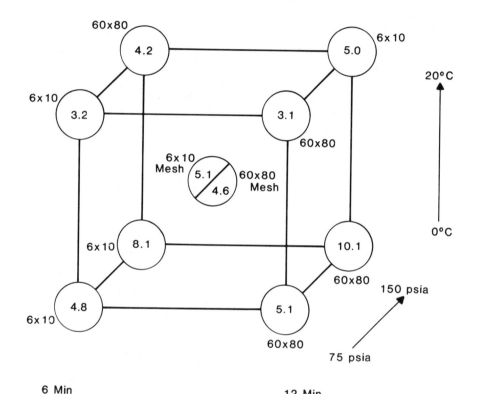

$$\text{Separation Factor} = \frac{N_2 \text{ (ads)}/CH_4 \text{ (Ads)}}{N_2 \text{ (Gas)}/CH_4 \text{ (Gas)}}$$

Figure 10. CH_4/N_2 separation factors for a 3 variable factorial design experimental series.   Feed was 33.3 SCC/g of zeolite per cycle of 40%N_2/60%CH_4 gas.

At the bottom of Figure 9, capacities derived from the clino N_2 adsorption isotherm (see Figure 3) are also listed. The factorial design results suggest that over 90% of the theoretical N_2 adsorption capacity can be dynamically used in a pressure swing process provided absorbent regeneration is adequate. Note that column temperature does not strongly affect dynamic N_2 adsorption capacity, but does influence N_2/CH_4 separation factors in the process. This results from a slowing of CH_4 adsorption kinetics as temperature is decreased.

The longest step in the pressure swing cycle for CH_4/N_2 over clinoptilolite is regeneration. Because this step involves vacuum and represents unproductive time in the process, regeneration was closely studied. Regeneration times were varied from 10 sec. to 20 minutes and regeneration mode was changed from vacuum (approx. 20 torr system pressure) to quiescent to CH_4 purge. The results for 5 of these runs are summarized in Figure 11 for a 20% N_2, 80% CH_4 feed gas. It appears that a vacuum regeneration time of at least 6 minutes is required to maintain product quality over multiple cycles. Other modes of regeneration were not successful and it appears that a process design cannot reasonably incorporate anything but vacuum regeneration.

Process Design & Economics

Using the detailed mass balance data for the experiments described above, a commercial process design was developed and subjected to an economic analysis. The process involves the use of 10 adsorber beds sized to handle 2.0 or 20.0 x 10^6 SCF/day of feed gas containing 20-30% N_2 and 70-80% CH_4. The product varies from 90-95% methane depending upon the feed composition. Pressure swing separations costs were found to be $.35/$10^6$ Btu for the larger plant and $1.20/$10^6$ Btu for the smaller plant. Corresponding costs for cryogenic separation of the same feedstocks were estimated to be $.29/$10^6$ Btu for the 20 x 10^6 SCF/day plant and $1.66/$10^6$ for the smaller plant. In all cases, capital charges assumed a 20% DCF-ROI. Thus pressure swing separation of CH_4/N_2 appears to be viable for small plants. Complete details of the process design and economic evaluation are being prepared for publication elsewhere.

Conclusions

The major points to be drawn from this work are:

The use of molecular sieving for the diffusive separation of molecules very close in size (3.64 Å vs 3.8 Å kinetic diameters) with commercially viable productivities is possible.

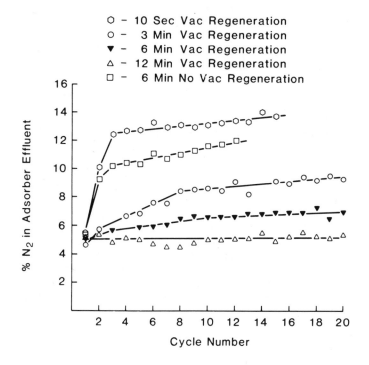

Figure 11. Product quality from an adsorber varies as the regeneration time is varied. Feed gas is $20\%N_2/80\%CH_4$.

The surface cation population of the clinoptilolite adsorbent controls the effectiveness of CH_4/N_2 separation.

The pressure swing separation of CH_4/N_2 is economically viable for small plant sizes, but cryogenic separation is more attractive at larger plant sizes.

Adsorbent improvement and engineering optimization may yield additional cost improvements.

Prior to an actual field trial of this technology, additional research is required. This additional work includes:

Determining the effect of trace inorganic gases (H_2O, CO_2, H_2S...) on the separation.

Determining the effect of higher hydrocarbons on long-term adsorbent performance.

Optimization of CH_4 and N_2 adsorption kinetics through the control of adsorbent surface cation population.

Engineering optimization, especially in the areas of adsorber sizing and aspect ratio as well as process flow strategy.

Acknowledgments

We wish to express our gratitude to Dr. J. Shih for his assistance in the economic analyses and to Mr. L.A. Young for his process control and data analysis programming for these experiments. We also thank Occidental Minerals, Inc. for partial funding of this research.

Literature Cited

(1) H.W. Habgood, U.S. Patent 2,843,219, 7/15/58.
(2) A. Alberti, TMPM Tschermaks, Min. Petr. Mitt. 22, 25-37 (1975).
(3) K. Koyama and Y. Takeuchi, Zeitschift fur Kristallographie 145, 216-239 (1977).
(4) A.B. Merkle and M. Slaughter, American Mineralogist 53, 1120-1138 (1968).
(5) N.Y. Chen, et al., Natural Zeolites, L.B. Sand and F.A. Mumpton, ed., Pergamon Press, N.Y., 1967, p. 411.
(6) S.A. Greene & H. Pust, J. of Physical Chemistry 65 (1), 55 (1958).
(7) P.E. Eberly, Jr., J. of Physical Chemistry 65 (1), 68 (1961).

(8) P.E. Eberly, and E.H. Spencer, Trans, Fara. Soc. 57, 289
     (1961).
(9) S. Ross, J.K. Saelens, and J.P. Olivier, J. of Physical
     Chemistry 66 (4), 696 (1962).
(10) R.J. Neddenriep, J. of Coll. & Interf. Science 28 (2), 293
     (1968).
(11) CRC Handbook for Chromatography, Vol. II., Gunter Zweig and
     J. Sharma, ed., CRC Press, West Palm Beach, Fla., 1972.
(12) Donald W. Breck, Zeolite Molecular Sieves, J. Wiley & Sons,
     New York, 1974.
(13) R.M. Barrer, Molecular Sieve Zeolites II, Advances in
     Chemistry Series No. 102, R.F. Gould, Ed., American Chemical
     Society, Washington, D.C., 1971.

RECEIVED December 27, 1982

Separation of Methane from Hydrogen and Carbon Monoxide by an Absorption/Stripping Process

VI-DUONG DANG

The Catholic University of America, Department of Chemical Engineering and Materials Science, Washington, DC 20064

An absorption/stripping process calculation using propane as the absorption solvent for separation of methane from hydrogen and carbon monoxide has been performed. Detailed material and energy balances for the process as well as the dimensions of the absorber and the stripper are reported. Other major pieces of equipment such as heat exchangers pumps, and compressors were evaluated in order to determine the equivalent electrical energy of the process as approximately 13550 cal(e)/gm-mole methane produced. The purity of methane in the final stream is 96% by volume at 100°F and 1000 psia. The present process appears to be a potential working process for methane separation in large quantity.

In coal gasification processes such as the flash hydropyrolysis or the Exxon catalytic process, methane is the major desired end product. However, methane is usually produced along with hydrogen and/or carbon monoxide. To obtain pipeline grade methane, it is necessary to develop economical methods of separating methane from hydrogen and/or carbon monoxide mixtures. The hydrogen and carbon monoxide are then recycled in the process. Several separation technologies such as absorption/stripping, cryogenic, clathrate formation have been examined (1). In this report, a method of absorption/stripping using propane as the solvent is presented. Absorption/stripping is a well-known chemical engineering operation and can be readily designed and constructed on a large scale. Therefore, the present method can be considered as a potentially useful method.

In the present work, a feed gas mixture of 40% CH_4, 45% H_2, and 15% CO at 100°F and 500 psia is to be separated. A production capacity of 250 x 10^6 scf/day of CH_4 is assumed. The methane is delivered to the pipeline at 100°F and 1000 psia. For the application of absorption/stripping, it is desirable to utilize a

0097–6156/83/0223–0235$06.00/0

solvent which has a high solubility for methane. Table I shows
some solubility data for methane in water and in various organic
solvents. It is seen that methane has a higher solubility in
propane than in any of the other solvents listed. Therefore, propane was chosen as the solvent for the absorption/stripping separation process investigated in this report.

Operating conditions for the absorption and stripping towers
are important design parameters for the process. Due to vapor
pressure and entrainment, propane will be present in the effluent
gas streams from both the absorber and stripper. Usually this
quantity of propane is not recovered and is considered an economic
loss. The amount of propane in the gas phase is mainly dependent
on the operating temperature and pressure of the towers.

Figure 1 shows the vapor pressure of some of the relevant
compounds as a function of temperature. At $-230.8\,°F$ and $-184\,°F$,
the vapor pressure of propane is about 0.1 mm Hg and 1 mm Hg
respectively. Absorption/stripping process is a conventionally
practical process so it is decided to evaluate the separation of
methane from hydrogen and carbon monoxide by this process. Since
the operating pressures of the absorber and stripper are about
500 psia and 487 psia respectively, the mole fractions of propane
in the outlet gas streams of the absorber and stripper are about
6.75×10^{-6} and 1×10^{-4} at $-230.8\,°F$ and $-184\,°F$ respectively.
When the absorber operates at higher temperature as shown in
Table II, propane losses increase. When the absorber and stripper
operate at $-230.8\,°F$ and $-184\,°F$ respectively, the propane loss is
equivalent to about 0.0850% and 3.28% respectively of the cost of
methane assuming the cost of methane and propane is the same.
This low percentage of propane loss was considered acceptable.
Hence it was determined to operate the absorber at $-230.8\,°F$ and
the stripper at $-184\,°F$.

Process Description

A schematic process flow sheet is shown in Figure 2. Inlet
gas, a mixture of methane, hydrogen and carbon monoxide at $100\,°F$
and 500 psia (stream 1) is successively cooled to $-140\,°C$ by the
outlet gas stream from the absorber and some recycle gas. The
absorber is a packed column of 1-in. berl saddles with 50% void
fraction. The rich liquid from the bottom of the absorber is heat
exchanged with the bottom liquid from the stripper. The stripper
is also a packed column with 1-in. berl saddles. The dissolved
methane, hydrogen, and carbon monoxide is stripped out by heating
at the bottom of the stripper. The outlet gas stream from the
stripper is heated in a heat exchanger by a recycle gas stream and
is further compressed to produce the final methane product at
$100\,°F$ and 1000 psia.

To perform a further detailed process calculation on this
multi-component absorption and stripping process, vapor liquid
equilibrium data for methane, hydrogen, and carbon monoxide is

Table I. The Solubility of Methane in Water and in Organic Solvents at 77°F and at Pressures up to 140 Atmospheres

Pressure, atm	cc CH_4 (at 77°F and 760 mm) dissolved in 1.0 cc solvent at pressure					
	20	40	60	80	100	140
Water	0.9	1.2	1.8	2.0	2.6	3.3
Methanol(1)	9.0	19.0	29.5	41.0	55.0	—
Ethanol	7.5	14.5	21.5	29.5	38.0	48.0 (120 atm)
Propane	41	83	125	174	200 (90 atm)	—
Butane	37	71	107	144	—	—
Pentane	33	63	97	129	164	—
Hexane	30	59	89	120	—	—
Octane	24	51	80	110	140	—
Cyclohexane	15	33	56	83	—	—
Benzene	11	24	41	59	80	—
Heavy Naphtha	10	20	31	43	—	—
Gas Oil	8	16	23	32	41	62

(1) The solubility of methane in n-propanol, isopropanol, n-butanol and iso-butanol is, within experimental error, the same as in methanol.

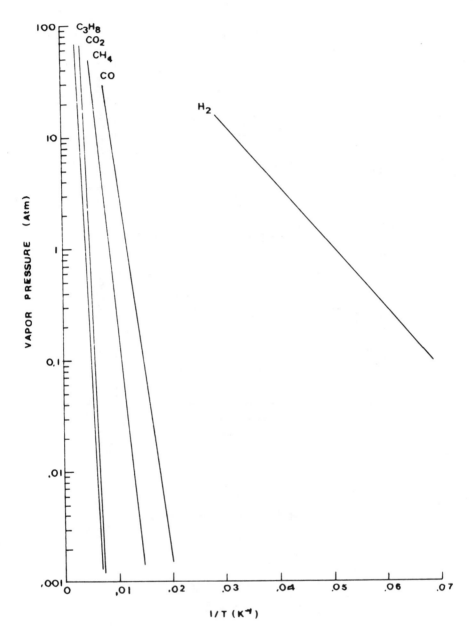

Figure 1.   Vapor pressure of selected compounds vs. 1/T.

Table II.   Propane Losses at Different Absorber Temperatures
(Stripper Temperature = −184°F)
Plant Capacity: 250x10^6 SCF/Day

T(°F)	lb/day loss of Propane			% of Propane Cost Loss Assuming Methane and Propane Cost to be the Same
	Absorber	Stripper	Absorber + Stripper	
− 58	2.79x10^6			
− 76	1.58x10^6			
−112	4.81x10^5			
−148	1.11x10^5	8415	1.19x10^5	1.133
−200	4.77x10^3	8415	1.319x10^4	0.126
−230.8	4.77x10^2	8415	8892	0.085

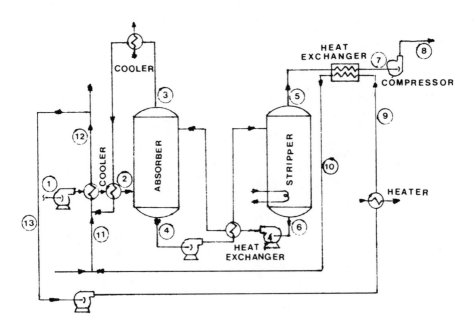

Figure 2.   Flow diagram for methane separation by absorption/stripping method.

needed. Equilibrium K-values of methane and propane are obtained
from DePriester (2). Solubility data for hydrogen in propane is
obtained from Trust and Kurata (3). Solubility of carbon monoxide
in propane is obtained from Prausnitz and Shair (4).

The absorber calculation is based on 1 lb-mole of gas stream
entering into the bottom of the absorber. The entering liquid is
fed at a rate of 1.5 lb-mole/lb mole of gas. For 95% methane
removed, a preliminary estimate of methane absorbed is 0.38 lb-
mole, carbon monoxide absorbed is 0.13 lb-mole, hydrogen absorbed
is 0.1 lb-mole. The absorber is operated under isothermal conditions of -146°C. The liquid to gas mass ratio at the top of the
tower is 3.85 while at the bottom it is 2.11 with an average value
of 2.98. The key component is methane. Using a method of average
absorption factor which for methane is 6.478 and 95% methane removal, it is possible to determine the number of theoretical transfer units to be 2. With these conditions, methane in the outlet
gas stream of the absorber is 0.002 lb-mole, carbon monoxide is
0.13 lb-mole and hydrogen 0.44 lb-mole. The molar concentrations
of the absorber effluent gas are CH_4, 3.5×10^{-3}; CO, 0.227; H_2,
0.769; C_3H_8, 6.75×10^{-6}. The molar concentrations in the liquid
phase are CH_4, 0.101; CO, 0.005; H_2, 0.003; C_3H_8, 0.891.

Absorber design requires the calculation of the quantity

$\dfrac{L}{G}\sqrt{\dfrac{\rho_G}{\rho_L}}$ which is 0.79 for the present case where L and G are mass

velocities of liquid and gas, ρ_L and ρ_G are densities of liquid
and gas. At flooding conditions

$$\frac{G^2\left(\dfrac{a_p}{\varepsilon^3}\right)(\mu_L^{'})^{0.2}}{g_c\rho_G\rho_L} = 0.05$$

where a_p is surface area of packing per unit tower volume, ε is
fractional void volume of dry packing $\mu_L^{'}$ is liquid viscosity, g_c
is local gravitational constant. Therefore assuming 60% of
flooding conditions, G is 2920 lb/hr ft^2. Since it is desired to
produce 250×10^6 scf/day methane, the inlet gas flow into the
absorber is 2.35×10^6 lb/hr.

From this and the gas mass velocity, it is determined that
3 towers are needed, each one with a diameter of 18.5 ft. The
tower height is determined by the height of a transfer unit which
is 20.7 ft. (5). Therefore, the height of the packed tower is
41.4 ft. Other detailed tower characteristics are given in
Table III.

The stripping tower is operated at -120°C and 487 psia and
the molar ratio of liquid to gas is 9.7. Again taking methane as
the key component, and 95% methane stripped, about 0.0962 lb mole

Table III. Summary of Dimensions and Characteristics
            of Absorbers and Strippers

		Absorbers
Diameter of Absorber	=	18.5 ft
Number of Absorber	=	3
Height of Packed Tower	=	41.4 ft
Pressure Drop in Packed Tower	=	1.94 psia
Wall Thickness of Tower	=	2.98 in.
		Strippers
Diameter of Stripper	=	17.6 ft
Number of Stripper	=	3
Height of Packed Tower	=	22.8 ft
Pressure Drop in Packed Tower	=	0.41 psia
Wall Thickness of Tower	=	2.35 in.

of methane, 0.0045 lb-mole of carbon monoxide, 0.002 lb-mole of
hydrogen are stripped. The preliminary estimates of the liquid to
gas ratio is 26.7 at the top and 24.0 at the bottom with an average
of 25.4. The average stripping factor for methane is 21.2 and two
transfer units are needed. Under these conditions 0.0046 lb-mole
of methane, 0.0032 lb-mole of carbon monoxide, and 0.00036 lb-mole
of hydrogen remain in the bottom liquid phase. In the bottom outlet stream of the liquid phase, mole fractions of the chemical
species are CH_4, 0.0051; CO, 0.0036; H_2, 0.0004; C_3H_8, 0.991. The
outlet mole fraction of the gas phase from the stripper is CH_4,
0.960; CO, 0.019; H_2, 0.021; C_3H_8, 0.0001.
    To design the stripper, again methane is chosen as the key
component.

$\dfrac{L}{G}\sqrt{\dfrac{\rho_G}{\rho_L}}$ is taken to be 7.81. At flooding

$$\frac{G^2 \left(\dfrac{a_p}{\epsilon^3} \right) (\mu_L')^{0.2}}{g_c \rho_g \rho_L} = 0.0018$$

Therefore, the gas mass velocity is 1372 lb/hr ft^2 for 60% flooding condition with a gas mass flow of 1.28 x 10^6 lb/hr. It is
determined that 3 towers are required each one 17.6 ft in diameter. The height of a transfer unit is a 11.4 ft and the total
height of a packed tower is 22.8 ft. Pressure drop in the tower
is 0.41 lb/in^2. A summary of the composition of various streams
is given in Table IV and stripper characteristics are given in
Table III.
    A thermal energy balance of the process is performed for the
coolers, for streams 1 and 2 in Figure 2 and the heat exchanger
between streams 5 and 7. The heat input for the heat exchanger
between the absorber and the stripper is considered to be negligible. The electrical power input to the compressor is 5135
horsepower. The thermal energy load for the heater between
streams 13 and 9 in Figure 2 is 2.2 x 10^7 cal/min and for the
cooler at the outlet gas stream from the absorber is 5.06 x 10^7
cal/min. A further breakdown of the thermal energy requirement
of the process is given in Table V. Detailed calculations are
given in the Appendix. Combining the thermal energy and the
pumping and compressing power required for the process as shown
in Figure 2, the total thermal energy required is 30955 cal/g-mole
which is about 11.8 times higher than the minimum ideal separation
energy required which is calculated to be 2621 cal/g-mole CH_4.

Table IV. Temperature, Pressure, Flow Rate, and Mole Fraction of Different Numbered Streams in the Flow Diagram

Stream Number	Temp °F	Pressure psia	Flow Rate		Mole Fraction			
			$\frac{lb}{day}$	$\left(\frac{ft^3}{day}\right)$	CH_4	H_2	CO	C_3H_8
1	100	500	5.64×10^7	(2.32×10^7)	0.4	0.45	0.15	0
2	-230.8	500	5.64×10^7	(9.5×10^6)	0.4	0.45	0.15	0
3	-230.8	487	2.23×10^7	(8.68×10^6)	0.0035	0.769	0.227	6.75×10^{-6}
4	-230.8	500	8.48×10^8		0.1013	0.0025	0.0051	0.891
5	-184	485	3.06×10^7		0.96	0.021	0.019	0.0001
6	-184	487	8.12×10^8		0.0051	0.0004	0.0036	0.999
7	100	485	3.06×10^7		0.96	0.021	0.019	0.0001
8	100	1000	3.06×10^7	(4.64×10^6)	0.96	0.021	0.019	0.0001
9	110	487	7.1×10^6		0.0035	0.769	0.227	6.75×10^{-6}
10	-110	487	7.1×10^6		0.0035	0.769	0.227	6.75×10^{-6}
11	-100	487	1.77×10^8		0.0035	0.769	0.227	6.75×10^{-6}
12	90	487	2×10^8		0.0035	0.769	0.227	6.75×10^{-6}
13	90	487	7.1×10^6		0.0035	0.769	0.227	6.75×10^{-6}

Table V. Energy Requirement for the Separation of Methane by Absorption/Stripping Using Propane

Electrical Energy Requirement

Compressor Power (Horsepower)	=	5135
Pumping Power Required from Bottom of Absorber to Top of Stripper (Horsepower)	=	2.78×10^5
Pumping Power Required from Bottom of Stripper to Top of Absorber (Horsepower)	=	3.3×10^5
Pumping Power of Inlet Gas Blower (Horsepower)	=	4973
Estimated Pumping Power for Fluid Transport (Horsepower)	=	1796
Total Pumping Power/g-mole methane produced cal(ϵ)/g-mole	=	12334
Equivalent Total Thermal Pumping Power/g-mole CH_4 cal(t)/g-mole	=	30834

Thermal Energy Requirement

Thermal Energy Supplied to Heater Between Streams 13 and 9 cal(t)/g-mole	=	36.5
Thermal Energy Load to Cooler in Absorber Outlet cal(t)/g-mole	=	84.1
Total Thermal and Equivalent Electrical Energy cal(t)/g-mole	=	30955

Acknowledgment

Part of the work was performed at Brookhaven National Laboratory and was acknowledged hereby.

Literature Cited

1. Dravo Corporation "Hydrogasification Gas Processing Studies"; Contract CPD-7209, February 27, 1978.
2. DePriester, C. L. "Light-Hydrocarbon Vapor-Liquid Distribution Coefficients, Pressure-Temperature-Composition Charts and Pressure-Temperature Nomographs"; Chem. Eng. Progress Symp. 1953, 7, 49.
3. Trust, D. B.; Kurata, F. "Vapor-Liquid Phase Behavior of the Hydrogen-Carbon Monoxide-Propane Systems"; AIChE J. 1971, 17, 86.
4. Prausnitz, J. M.; Shair, F. H. "A Thermodynamic Correlation of Gas Solubilities"; AIChE J. 1961, 7, 682.
5. Bennett, C. O.; Myers, J. E. "Momentum, Heat and Mass Transfer"; 2nd ed., McGraw-Hill Co., 1974, pp. 571-574.

Appendix

Energy Calculation for the Separation of Methane from Process Gasification Stream Using Propane Absorption/Stripping

The energy calculations for the process are based on electrical and thermal energy. It is assumed that the heat exchanger between the absorber and the stripper operates ideally so no net heat loss or gain will be realized in this unit. The electrical energy requirements are for (1) the pumps between the absorber and the stripper, (2) the inlet gas blower, (3) the compressor at the end for the product methane, and (4) pumping power for fluid transport. Thermal energy is supplied for (1) the heater between streams 9 and 13, and (2) the cooler at the outlet of the absorber. The calculations will be performed separately in the following sections.

A. Pumping Power Required to Transport Fluid from Bottom of Absorber to Top of Stripper

Liquid flow rate $= 8.48 \times 10^8$ lb/day

$$= \frac{8.48 \times 10^8 \text{ lb/day} \times 7.48 \text{ gal/ft}^3}{36.52 \text{ lb/ft}^3 \times 24 \text{ hr/day}}$$

$$= 7.24 \times 10^6 \text{ gal/hr} = 1.21 \times 10^5 \text{ gal/min}$$

Density of liquid $= 36.52$ lb/ft^3

Height of stripping tower packing is 22.8 ft. Consider heading of the tower is 20% additional. Then the tower height ΔZ is 27.4 ft. The fluid is transported by 30 pipes and the diameter of the pipe is about 0.25 ft. The energy balance for each pipe is

$$\frac{g}{g_c} \Delta Z + 0.8 \, W_s + \frac{2fu^2 L}{g_c D} = 0 \tag{A-1}$$

Velocity u in each pipe $= \dfrac{8.48 \times 10^8 \times 4}{24 \times 36.52 \times \pi \times 30 \times .25^2}$

$$= 6.56 \times 10^5 \text{ ft/hr} = 182.4 \text{ ft/sec}$$

$$Re = \frac{uD\rho}{\mu} = \frac{182.4 \text{ ft/sec} \times 0.25 \text{ ft} \times 36.52 \text{ lb/ft}^3}{1.2 \times 10^{-2} \times \frac{36.52}{62.4} \times 6.72 \times 10^{-4}} = 3.53 \times 10^8$$

Relative roughness factor and friction factor are taken to be 0.0006 and 0.004, respectively. Therefore, the frictional loss term is

$$1w_f = \frac{2 \times 0.004 \times 182.4^2 \times 37.4}{32 \times 0.25} = 1244 \text{ ft } 1b_f/1b$$

From Eq. (A-1), we can calculate the loss per set of pipes (10 pipes) connecting each absorber and stripper is

$$W_s = \frac{-(1244 \times 10 + 27.4)}{0.8} = 15584 \text{ ft } 1b_f/1b$$

The total electrical power required for the 30 pipes is

$$\mathbb{P} = \frac{15584 \text{ ft-}1b_f/1b \times 8.48 \times 10^8 \text{ 1b/day} \times 5.05 \times 10^{-7} \text{ } \mathbb{P} \text{ hr/ft-}1b_f}{24 \text{ hr/day}}$$

$$= 2.78 \times 10^5 \text{ } \mathbb{P}$$

B. **Pumping Power Required to Transport Fluid from Bottom of Stripper to Top of Absorber**

Liquid flow rate $= 8.12 \times 10^8$ 1b/day

$$= \frac{8.12 \times 10^8 \text{ 1b/day} \times 7.48 \text{ gal/ft}^3}{36.52 \text{ 1b/ft}^3 \times 24 \text{ hr/day} \times 60 \text{ min/hr}}$$

$$= 1.15 \times 10^5 \text{ gal/min}$$

Again we take 20% additional height for the height of the absorber which has a packing height of 41.4 ft so the total height of the absorber is 50 ft. Let's apply the same mechanical energy balance Eq. (A-1) to calculate the pumping power here. Velocity of the fluid is

$$U = \frac{8.12 \times 10^8 \times 4}{24 \times 36.52 \times \pi \times 30 \times .25^2} = 6.29 \times 10^5 \text{ ft/hr} = 174.9 \text{ ft/sec}$$

$$Re = \frac{uD\rho}{\mu} = \frac{174.9 \times 0.75 \times 36.52 \text{ 1b/ft}^3}{1.2 \times 10^{-2} \times \frac{36.52}{62.4} \times 6.72 \times 10^{-4}} = 1.02 \times 10^9$$

Using the relative roughness factor and friction factor to be 0.0006 and 0.004 again. The frictional loss in the pipe is

$$1w_f = \frac{2 \times 0.004 \times 174.9^2 \times 50}{32.2 \times .25} = 1520 \text{ ft-lb}_f/\text{lb}$$

So the total energy required per set of pipe (10 pipes) connecting each stripper and absorber is

$$W_s = \frac{-(1520 \times 10 + 50)}{0.8} = 19063 \text{ ft-lb}_f/\text{lb}$$

The total pumping power is

$$P = \frac{19063 \text{ ft-lb}_f/\text{lb} \times 8.12 \times 10^8 \text{ lb/day} \times 5.05 \times 10^{-7} \text{ P hr/ft-lb}_f}{24 \text{ hr/day}}$$

$$P = 3.3 \times 10^5$$

C. **Inlet Gas Blower**

Incoming gas to absorber is $2.32 \times 10^7 \text{ ft}^3/\text{day}$. Main flow rate at absorber is 2920 lb/hr ft^2 at a density

$$\rho_g = \frac{PM_{av}}{Z_{av}RT}, \quad T = 230°F = 127°K$$

$$CH_4: \quad T_r = \frac{127}{191} = 0.665, \quad P_r = \frac{500}{14.7 \times 45.8} = 0.7426, \quad Z_c = 0.56$$

$$CO: \quad T_r = \frac{127}{133} = 0.955, \quad P_r = \frac{500}{14.7 \times 34.5} = 0.986, \quad Z_c = 0.32$$

$$H_2: \quad T_r = \frac{127}{33.3} = 3.814, \quad P_r = \frac{500}{14.7 \times 12.8} = 2.657, \quad Z_c = 0.98$$

$$Z_{av} = 0.4 \times 0.56 + 0.15 \times 0.32 + 0.45 \times 0.98 = 0.71$$

$$M_{av} = 0.4 \times 16 + 0.15 \times 28 + 0.45 \times 2 = 11.50$$

$$\rho_g = \frac{500 \times 11.5}{0.71 \times 10.73 \times 230} = 3.28 \text{ lb/ft}^3$$

$$\frac{G}{\rho_g} = \frac{2970 \text{ lb/hr ft}^2}{3.28 \text{ lb/ft}^3} = 905.5 \text{ ft/hr} = 0.25 \text{ ft/sec}$$

Number of tubes in absorber = n

$$U = \frac{2.32 \times 10^7 \ ft^3/day \times 4}{24 \ hr/day \times \pi \times nD^2 \ ft^2} = \frac{1.231 \times 10^6}{nD^2} \ ft/hr = \left(\frac{342}{nD^2}\right) ft/sec$$

$$Re = \frac{uD\rho}{\mu} = \frac{342}{nD^2} \times \frac{D \times 3.28 \ 1b/ft^3}{0.007^{cP} \times 6.72 \times 10^{-4}} = \frac{2.38 \times 10^8}{nD}$$

$$1w_f = \frac{2fLU^2}{g_c D} = \frac{2 \times 0.004 \times 10 \ ft}{32.2 \ D} \times \frac{342^2 \ \frac{ft^2}{sec^2}}{nD^2} = \frac{291}{n^2 D^5} \ \frac{ft-1b_f 1b}{}$$

let n = 30, D = 0.25 ft

Frictional loss/tube = 330.63 ft-1b$_f$/1b

$$\frac{g}{g_c} \Delta Z + 1w_f + \eta W_s = 0 \qquad\qquad (A-2)$$

gas flow = $2.32 \times 10^7 \ ft^3/day$ (=5.64×10^7 1b/day)

Pressure drop in tower = 1.3 in H$_2$0/ft depth, ΔZ = 41.4 ft

$$\text{Total pressure drop} = \frac{1.3 \times 41.4 \ ft \ H_2 0}{12} \ \frac{g}{g_c} = 4.49 \ ft-1b_f/1b$$

Applying Eq. (A-2) to one set of pipe (10 pipes) for incoming gas into the absorber, one gets

$$(41.4+4.49) + 330.63 \times 10 + \eta W_s = 0$$

$$W_s = -\frac{4.49+41.4+3306.3}{0.8} = -4190.2 \ ft-1b_f 1b$$

$$\text{Energy required} = \frac{5.64 \times 10^7 \ \frac{1b}{sec}}{24 \ hr/day \times 3600 \ sec/hr} \times \frac{4190.2}{550} = 4973 \ \mathbb{P}$$

D. Compressor at End of Product Methane Between Streams 7 and 8

$$\mathbb{P} = \frac{3.03 \times 10^{-5}}{k-1} P_1 \ q_{f_{m_1}} \left[\left(\frac{P_2}{P_1}\right)^{\frac{k-1}{k}} -1 \right] \qquad (A-3)$$

where
$$k = \frac{C_p}{C_v} = 1.31 \text{ for methane}$$

For the present case

$$\mathbb{P} = \frac{3.03 \times 10^{-5}}{1.31-1} \times 485 \text{ lb/in}^2 \times 144 \text{ in}^2/\text{ft}^2$$

$$\times \frac{4.64 \times 10^6 \text{ ft}^3/\text{day}}{24 \text{ hr/day} \times 60 \text{ min/hr}} \left[\left(\frac{1000}{485}\right)^{\frac{.31}{1.31}} - 1 \right]$$

$$\mathbb{P} = 4108 \ \mathbb{P}$$

If the compressor is 80% efficient, then the power required = 5135 $\mathbb{P}$.

E. Pumping Power for Fluid Transport

Pumping power is required to transport fluid in the streams 9, 10, and 13. Gas flow rate and density in these streams are 7.1×10^6 lb/day and 0.44 lb/ft^3, respectively. Using a 0.25 ft diameter pipe, velocity in the pipe is

$$\frac{7.1 \times 10^6 \text{ lb/day} \times 4}{30 \times 0.44 \text{ lb/ft}^3 \times 24 \times 3600 \times \pi \times .25^2} = 127 \text{ ft/sec}$$

$$Re = \frac{uD\rho}{\mu} = \frac{127 \times 2.5 \times 0.44}{(.0183 \times .23 + .009 \times .77) \times \frac{.44}{62.4} \times 6.72 \times 10^{-4}} = 2.65 \times 10^8$$

Relative roughness factor and friction factor are again taken to be 0.0006 and 0.004, respectively. Assuming the total transport line as 600 ft, frictional loss in fluid transport is

$$1w_f = \frac{2fu^2L}{g_c D} = \frac{2 \times 0.004 \times 127^2 \times 600}{32.2 \times 0.25} = 9617 \text{ ft-lb}_f/\text{lb}$$

With 80% efficiency of the pump, total energy loss is 1.2×10^4 ft-lb$_f$/lb and the pumping power required is

$$\mathbb{P} = \frac{1.2 \times 10^4 \text{ ft-lb}_f/\text{lb} \times 7.1 \times 10^6 \text{ lb/day} \times 5.05 \times 10^{-7} \ \mathbb{P} \text{-hr/ft-lb}_f}{24 \text{ hr/day}}$$

$$= 1796 \ \mathbb{P}$$

F. Thermal Energy Required in Heat Exchanger Between Streams 13 and 9

$Q = 7.1 \times 10^6$ lb/day x 7 Btu/lb-mole °F $\big/$ 7.89 lb/lb-mole (110-90) °F

 $= 1.13 \times 10^8$ Btu/day

 $= 1.3 \times 10^8$ Btu/day x 252 cal/Btu $\big/$ 24 hr/day x 60 min/hr

 $= 2.2 \times 10^7$ cal/min

Thermal energy supplied to the heat exchanger

$$= \frac{2.2 \times 10^7 \text{ cal/min x 24 hr/day x 60 min/hr x 16 g/g-mole}}{3.06 \times 10^7 \text{ lb/day x 454 g/lb}}$$
$$= 36.5 \text{ cal(t)/g-mole}$$

G. Energy Required to the Cooler in Absorber Outlet Stream

$(-Q) = 2.23 \times 10^7$ lb/day x 7 Btu/lb-mole °F $\big/$ 9.73 lb/lb-mole

  (-156 + 146) x 1.8

 $= 2.89 \times 10^8$ Btu/day

 $= 2.89 \times 10^8$ Btu/day x 252 cal/Btu $\big/$ 24 hr/day x 60 min/hr

 $= 5.06 \times 10^7$ cal/min

Energy supplied to the cooler

$$= \frac{5.06 \times 10^7 \text{ cal/min x 24 hr/day x 60 min/hr x 16 g/g-mole}}{3.06 \times 10^7 \text{ lb/day x 454 g/lb}}$$

It is also interesting to report the theoretical minimum energy requirement for the present process

$$W_{min,T} = RT \sum_j x_j F \ln \frac{P_2}{x_{jF} P_1}$$

For the present process, composition temperature and pressure of incoming gas are: CH_4, 0.4; H_2, 0.45; CO, 0.15; 100°F and 500 psia, respectively. Temperature and pressure of final product gases are at 100°F and 1000 psia. For complete separation

$$W_{min,T} = 1.98 \text{ Btu/lb-mole } °Rx560°R \left[0.4 \ln \frac{1000}{0.4x500} \right.$$

$$+ 0.45 \ln \frac{1000}{.45x500} + 0.15 \ln \frac{1000}{.15x500} \right]$$

$$= 1889 \text{ Btu/lb-mole x } 252 \text{ cal/Btu} \Big/ 454 \text{ g-mole/lb-mole}$$

$$= 1049 \text{ cal/mole gas} \Big/ 0.4 \frac{\text{mole } CH_4}{\text{mole gas}} = 2621 \frac{\text{cal}}{\text{mole } CH_4}$$

RECEIVED December 29, 1982

High-Temperature Hydrogen Sulfide Removal Using a Regenerable Iron Oxide Sorbent

S. S. TAMHANKAR and C. Y. WEN

West Virginia University, Department of Chemical Engineering, Morgantown, WV 26506

A state-of-the-art review is presented of high temperature H_2S removal and sorbent regeneration using an iron oxide sorbent, with a particular emphasis on kinetic and mechanistic studies of the reactions involved. Selection criteria for a high temperature H_2S sorbent and the various sorbent regeneration options are first briefly discussed. Results are then presented on the kinetics of the various reactions involved in the absorption and the regeneration steps. These studies were conducted in a thermogravimetric analyzer (TGA). The weight change data obtained in the TGA was used in conjunction with x-ray and Mossbauer spectroscopic analyses of the solid samples to elucidate the mechanisms of the reactions. Conversion correlations are reported based on the grain model, which can be used for reactor design and for predicting reactor performance.

Significant effort is underway in the United States to develop and commercialize coal gasification processes for producing gaseous fuels. One of the major obstacles in the development of such a process is the presence of undesirable contaminants in the product gas stream. The major contaminant in coal gasification is hydrogen sulfide (H_2S), which is toxic, poisonous to downstream catalysts and extremely corrosive in nature. Control of H_2S in the fuel gas to a safe level is therefore essential. The H_2S removal requirements are even more critical when the fuel gas is used in combined cycle power generation or in fuel cells.

Conventional methods of H_2S removal use a cold scrubbing technique, wherein the gas needs to be cooled to near room temperature. Although such techniques are effective, they are associated with a loss of sensible heat of the gas. The other

disadvantages of cold scrubbing include restriction by availabil-
ity of water, sludge disposal problem and condensation and mixing of tars with the scrubber water posing handling problems. In
view of these, and mainly for thermal efficiency, removal of H_2S
at high temperatures is very attractive. Besides, fuel cells
and combined cycle power generation systems use the hot fuel gas
directly. This further improves the overall thermal efficiency.
Figure 1 shows a schematic of the integrated system configurations for various applications. The figure also includes the
limit values of H_2S concentration for these applications. The
subject of this paper is the bulk sulfur removal, whereby the
H_2S concentration is brought down to about 200 ppm.

Sorbent Selection

The criteria used in selecting a suitable sorbent are:
efficiency, sulfur capture capacity, kinetics, cost, physical
strength and regenerability. Sorbent efficiency implies the
maximum amount of H_2S that can be removed under a given set of
operating conditions, based on the thermodynamic equilibrium of
the system. Figure 2 (1) shows equilibrium H_2S conversions in
the presence of various sorbent materials in relation to the EPA
sulfur emission standards. At all the temperatures of interest
iron-based sorbents seem to be satisfactory. Zinc oxide is
probably the best in this respect, and can in fact be used for
the final polishing step to reduce H_2S level to less than 1 ppm.
Sulfur capture capacity of a sorbent depends on the stoichiometry
of the reaction. In the temperature range of interest, iron
oxide (Fe_2O_3) captures about 44 weight % sulfur.
An important asset of iron oxide as an H_2S sorbent is its
reactivity. The results of a comparative study shown in Figure
3 (1) reveal that iron oxide is by far the most active among
the oxides considered. This is an important factor in the
choice of Fe_2O_3 as a sorbent.
Based on criteria of efficient H_2S removal above 1000°F,
physical strength, regenerability, sorbent life and economic
feasibility, laboratory tests (2,3) led to the selection of
iron oxide mixed with fly ash, in extruded pellets, as a
promising sulfur-removal sorbent.
In a commercial scale operation physical strength of
the sorbent pellets is important, especially if a moving or a
fluidized-bed reactor is used. Pellets of iron oxide alone
are not strong enough; addition of a support material is
necessary. In so doing, the percentage of iron oxide should
be kept as high as possible to maintain the overall high
efficiency of H_2S removal. Extensive tests were conducted at
METC (4), and the best support and an optimum composition
with it were found. This sorbent consisted of 45 weight %

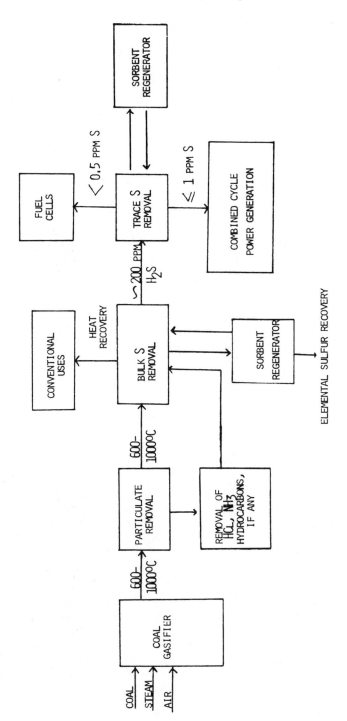

Figure 1. A schematic of an integrated coal gasification system for fuel cell or combined-cycle-power-generation applications.

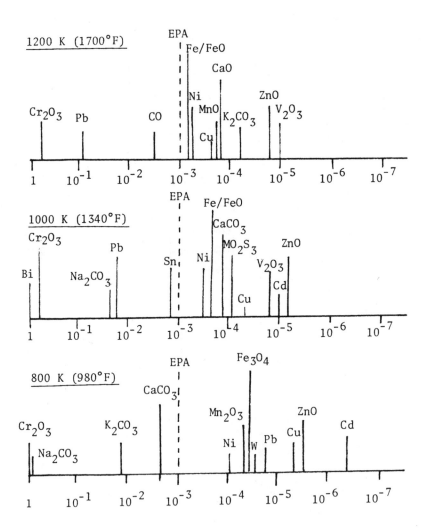

Figure 2. Equilibrium exit gas concentration of H_2S.

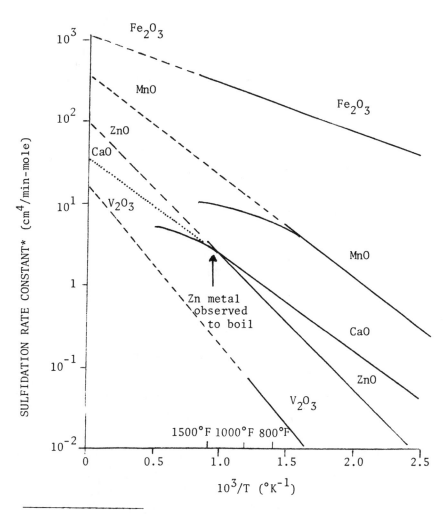

*Rates expressed per constant unit surfacd area

Figure 3. Kinetics of H_2S sorption by selected metal oxides.

Fe_2O_3 and 55 weight % silica, as support, with 1% sodium silicate as a binder. In the preliminary work fly ash was considered as a support material; but the maximum amount of iron oxide which could be incorporated with it was 25 weight %. Based on this and the other criteria discussed above, the iron oxide-silica composition was found to be the most suitable sorbent for bulk H_2S removal.

A simple method of regeneration of the sulfided sorbent is by its reaction with oxygen in air. With most metal sulfides this reaction results in sulfates at the temperatures of interest. Ferric sulfate has the lowest limit temperature of stability as shown in Table I (1). Hence, iron oxide is directly formed by regeneration in air. A major obstacle in

Table I
Final Product of Regeneration in Air
for Different Sorbents (Taken from Ref. 1)

Sulfide	→	Product	T(K)	Sulfide	→	Product	T(K)
FeS	→	$Fe_2(SO_4)_3$	< 950	CaS	→	$CaSO_4$	All
FeS	→	Fe_2O_3	<1680	NiS	→	$NiSO_4$	<1200
Cu_2S	→	$CuSO_4$	<1050	NiS	→	NiO	>1200
Cu_2S	→	CuO	<1300	ZnS	→	$ZnSO_4$	<1150
Cu_2S	→	Cu_2O	>1300	ZnS	→	ZnO	>1150
MnS	→	$MnSO_4$	<1100	MoS_2	→	MoO_3	<1350
MnS	→	Mn_2O_3	<1400	MoS_2	→	MoO_2	>1350
MnS	→	Mn_3O_4	>1400	CoS	→	CoO	>1000

the commercial development of a hot-gas desulfurization process has been the sorbent regeneration step. An efficient and economical method has not yet been established. Regeneration is discussed in more details in the next section.

As mentioned earlier, thermodynamically H_2S removal efficiency of ZnO is higher than that of Fe_2O_3. ZnO is therefore considered for the second stage, wherein H_2S concentrations can be reduced from about 200 ppm to less than 1 ppm, as required for fuel cell or combined cycle power generation applications. A further step in the development of a hot-gas desulfurization process is the use of a mixture of iron and zinc oxides. This novel concept, emerged at METC, uses zinc ferrite formed from a mixture of Fe_2O_3 and ZnO, and is aimed at reducing H_2S concentration to the desired level in a single step. The results so far (5) indicate that the lowest H_2S breakthrough concentration attained in this system is 10 ppm at a temperature of $1000^{\circ}F$. This approach appears attractive, but additional work is necessary before a definite conclusion can be drawn.

Regeneration Options

An important consideration in a hot-gas desulfurization process is an efficient and economical method of regeneration of the spent sorbent. This requires that the sulfur in the spent sorbent be recovered in a useful form. As mentioned earlier, air-oxidation is a convenient way of regenerating the sulfided sorbent to its original form. In this process the sulfur in the sorbent is converted to sulfur dioxide (SO_2). The SO_2 thus formed can then be used to make H_2SO_4, can be treated with limestone, or can be converted to elemental sulfur. The conversion to elemental sulfur can be achieved either by the Claus process or by treating SO_2 with a reductant such as carbon (Trail/Resox process). An attractive way of recovering sulfur is by reacting the SO_2 with iron sulfide itself to form iron oxide and elemental sulfur. Thus, if the SO_2 formed by the oxidation of iron sulfide is recirculated, elemental sulfur can be produced in a single step. The problem with this scheme is that the elemental sulfur needs to be separated from the exit stream by cooling, and the SO_2 needs to be reheated. Karr, et al. (6) and Schrodt and Best (7) carried out experiments to test the feasibility of this scheme. The latter authors, using coal ash as the sorbent material, concluded that sulfur recovery by this method is both technically and economically unattractive. The same may not be true for the iron oxide-silica sorbent. More studies are necessary before a definite conclusion can be drawn.

One of the problems with regeneration by air-oxidation is the large heat of reaction, which makes temperature control difficult. In the SO_2 recycle scheme this problem can be alleviated since the reaction of iron sulfide with SO_2 is endothermic. Another option being considered more recently for moderating the temperatures is to mix steam with air. In an exploratory study done at Air Products and Chemicals, Inc. (APCI) (8) some interesting observations were made. With very high steam concentrations (>90%) in the steam-air mixture, significant quantities of elemental sulfur were formed. This again presents an interesting possibility of regenerating the spent sorbent and forming elemental sulfur in a single step. Further studies are necessary to enable assessment of this process. Kinetic studies on this system are in progress at present at West Virginia University and will be discussed later in this paper.

The overall H_2S removal process would consist of two stages: the absorption stage, in which the sorbent is converted to the sulfide form by its reaction with the H_2S in the feed gas, and the regeneration stage, wherein the sulfide is converted to an oxide. It thus represents a cyclic process. Fixed-bed, moving-bed and fluidized-bed operations have been considered for the purpose. Kinetic data are necessary for choosing a proper reactor type and for the design

of a reactor. In the remainder of this paper attention will
be focused on this aspect. Some of the reactions involved
in this system have been studied before in different contexts.
However, the forms of reactants and the operating conditions
used were different. It is important to study these reactions
starting with the actual iron oxide-silica sorbent and under
conditions close to the real situation.

Reaction Mechanisms and Kinetic Studies

The objectives of the studies at West Virginia University
(WVU) were to identify the important reactions involved, to
elucidate the mechanisms and to study their kinetics. Experiments were conducted in a thermogravimetric analyzer (TGA)
apparatus; a general schematic of this is shown in Figure 4.
In this method, a change in the weight of the solid sample
during a reaction is continuously recorded as a function of
time. If the stoichiometry and the reaction mechanism are
determined, this weight change can be related to the solid
conversion. The conversion-time data can then be correlated
using an appropriate reaction model. In the following, important reactions, their mechanisms and the kinetic experiments
are described for each of the steps, viz. H_2S absorption,
sorbent regeneration by SO_2 recycle and regeneration by a
steam-air mixture. The discussion on the first two steps is
based on the previous publications (9,10) by the authors, which
provide the details.

H_2S Absorption Stage. In order to identify reactions that
would be important in this stage, reactions of the iron oxide
sorbent were carried out in the TGA using different synthetic
gas mixtures. A few important results of this study are shown
in Figure 5. Curve A in this figure represents the actual
reaction of Fe_2O_3 with a simulated gasifier product gas
(containing N_2, CO, CO_2, H_2 and H_2S). As may be noted, the
sample weight first rapidly decreased, reached a minimum and
then slowly increased again. Two distinct reactions were
suspected to be taking place. Hence, the sorbent was first
reacted with a mixture of CO and H_2, in the absence of H_2S.
This reduction reaction, which is very rapid, is represented
by curve B. The resulting sample was then reacted with H_2S
to obtain curve C. An interesting observation made is that
curve A can be obtained by adding curves B and C. Curve D
is for the reaction of Fe_2O_3 with H_2S in the absence of CO
and H_2. Obviously, this reaction is much slower than any
other reaction. Based on these observations it was concluded
that the important reactions in the absorber are: the reduction
of Fe_2O_3 to Fe and the subsequent sulfidation of Fe by H_2S,
which is rate-controlling. In addition, if CO_2 or H_2O are
present in the feed gas they would only affect the equilibirum
of the reduction reactions, which can be accounted for based on
the reported equilibrium data.

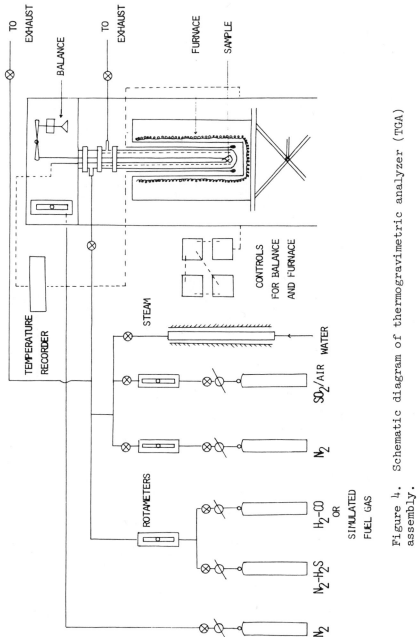

Figure 4. Schematic diagram of thermogravimetric analyzer (TGA) assembly.

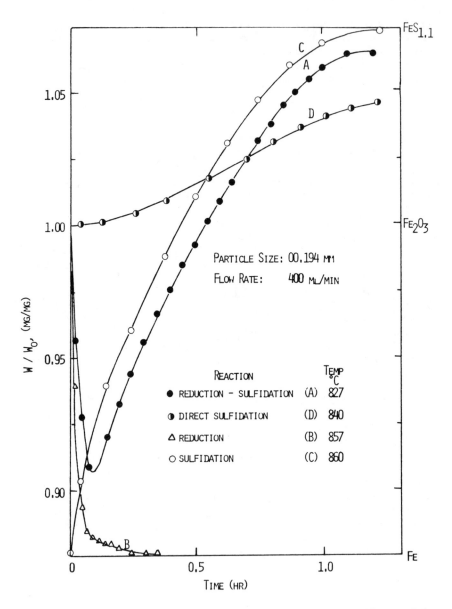

Figure 5. Comparison of weight change curves for reactions of iron oxide in different gas atmospheres.

The weight change data indicated that the reaction of Fe with H_2S results in a nonstoichiometric compound with the formula $FeS_{1.1}$. Mossbauer spectroscopic analysis of the samples confirmed this. The compound represents a pyrrhotite, a nonstoichiometric iron sulfide, which is well known in the Fe-S system. Based on the above observations, the reactions in an absorber can be written as follows:

$$Fe_2O_3 + 3/2(H_2 + CO) \rightleftharpoons 2Fe + 3/2(H_2O + CO_2) \quad (1)$$

$$\text{(for a fixed ratio of } H_2:CO)$$

and

$$Fe + 1.1\ H_2S \longrightarrow FeS_{1.1} + 1.1\ H_2 \quad (2)$$

Thus, reactions considered here are those taking place under actual process conditions. Reactions (1) and (2) were studied separately in the TGA at different temperatures and with different particle sizes. Also, effect of the gas-phase reactant concentration was studied to evaluate the reaction order. In a few experiments the effects of the presence of H_2O in the feed gas on reaction (1) and that of H_2 on reaction (2) were investigated.

Regeneration by SO2 Recycle Method. In this case, clearly, the two reactions, which need to be studied, are reactions of iron sulfide, $FeS_{1.1}$ with O_2 and with SO_2. The reaction of $FeS_{1.1}$ with O_2 was found to be very fast and formed Fe_2O_3 directly. No intermediates could be identified using Mossbauer spectroscopy.

The reaction of $FeS_{1.1}$ with SO_2 was rather interesting. In this reaction, $FeS_{1.1}$ was first converted to magnetite, Fe_3O_4, which was then slowly oxidized to Fe_2O_3. This is shown in Figure 6, where the minima in the weight change curves correspond to Fe_3O_4. This was also confirmed by x-ray and Mossbauer spectroscopic analyses. In fact it was found that up to the minimum point the only product formed was Fe_3O_4. These results are presented in Table II. Since the oxidation to Fe_2O_3 is relatively very slow only the first step was considered for the purpose here. Accordingly, the reactions in the regenerator would be

$$2FeS_{1.1} + 3.7\ O_2 \longrightarrow Fe_2O_3 + 2.2\ SO_2 \quad (3)$$

and

$$3FeS_{1.1} + 2\ SO_2 \rightleftharpoons Fe_3O_4 + 1.55\ S_2 \quad (4)$$

These two reactions were studied in the TGA under different operating conditions to obtain their kinetics. Reaction (3) is highly exothermic. Hence, low oxygen concentrations and

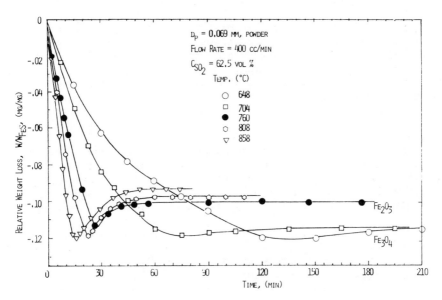

Figure 6. Effect of temperature on the weight loss of sulfided iron oxide sorbents in sulfur dioxide atmosphere.

Table II

Experimental Conditions and the Results of the Mossbauer Analysis
of Specimens Partially Reacted in SO_2 (1.5 x 6.0 mm pellets)

Run No.	Experimental conditions			atomic ratio	Results of the Mossbauer analysis conversion of $FeS_{1.1}$, X	
	C_{SO_2} (vol %)	T (°C)	Time (min)		based on weight change	based on Mossbauer analysis
815	—	—	0.0	$\dfrac{Fe_3O_4}{FeS_{1.1}} = 0.00$	0.0	0.0
817-3	62.5	759	45.0	$\dfrac{Fe_3O_4}{FeS_{1.1}} = 4.00$	0.572	0.555
817-2	62.5	760	60.0	$\dfrac{Fe_3O_4}{FeS_{1.1}} = 6.66$	0.690	0.635
818-2	62.5	759	97.5	$\dfrac{Fe_3O_4}{FeS_{1.1}} = 7.00$	0.700	0.670

Reprinted with permission from Ref. 10. Copyright 1981, Pergamon Press, Ltd.

high gas flow rates were used to maintain near-isothermal conditions.

   Regeneration by Steam-Air Mixtures. As mentioned earlier, preliminary exploratory work on this system was done at APCI. At APCI experiments were conducted in a fixed-bed reactor and product gas compositions were monitored. Typical results of this study are given in Table III ($\underline{8}$). According to APCI equilibrium calculations very little elemental sulfur yields were predicted. However, high sulfur yields were obtained, particularly with high steam concentrations. This was an interesting observation, and detailed studies on this system were warranted. As before, the objectives of the studies at WVU are to identify the reactions, to elucidate the mechanisms and then to study their kinetics.

   Based on the APCI results, a number of possible reactions can be postulated for the system iron sulfide - H_2O - O_2. These are listed below.

$$FeS_x + H_2O \rightleftharpoons FeO + H_2S \tag{5}$$

$$3FeO + H_2O \rightleftharpoons Fe_3O_4 + H_2 \tag{6}$$

$$2FeS_x + 7/2\ O_2 \longrightarrow Fe_2O_3 + 2SO_2 \tag{7}$$

$$3FeS_x + 2SO_2 \rightleftharpoons Fe_3O_4 + 5/2\ S_2 \tag{8}$$

$$H_2S + 1/2\ O_2 \rightleftharpoons H_2O + S_x \tag{9}$$

$$1/2\ SO_2 + H_2S \rightleftharpoons H_2O + S_x \tag{10}$$

$$S_x + O_2 \longrightarrow SO_2 \tag{11}$$

$$H_2S \rightleftharpoons H_2 + 1/2\ S_2 \tag{12}$$

$$1/2\ SO_2 + H_2 \rightleftharpoons H_2O + 1/4\ S_2 \tag{13}$$

Reactions (7) and (8) are the same as reactions (3) and (4) identified in the previous regeneration scheme, while reactions (9) and (10) represent the Claus process. Reaction (12) is postulated here to be a major reaction responsible for the elemental sulfur formation, particularly in the absence of air. Kinetic and thermodynamic equilibrium data for the catalytic decomposition of H_2S have been reported in the literature ($\underline{11},\underline{12}$).

   To delineate the various possible reactions, in this study reaction of $FeS_{1.1}$ with steam alone was investigated first. Experiments were carried out in the TGA using a steam-nitrogen mixture with the steam concentration of 94 percent by volume.

Table III

Effect of Water Content on Regeneration Product-Distribution
(Taken from the APCI Final Report, Ref. 8)

Run No.	2102	2118	2129
Regeneration Feed Gas, Mol Percent			
Water	85.0	95.0	95.0
Air	15.0	5.0	0.0
Sulfur Dioxide	0.0	0.0	0.0
Nitrogen	0.0	0.0	5.0
Sorbent	42 Percent Added Commercial Iron Oxide on Silica	42 Percent Added Commercial Iron Oxide on Silica	42 Percent Added Commercial Iron Oxide on Silica
Sorbent Volume, cc	40.0	40.0	40.0
GHSV	600.0	600.0	1200.0
Regeneration Duration, Hours	14.0	42.0	10.0
Reactor Block Temperature, $°F$	1000.0	1000.0	1000.0
Material Balance, Percent	95.0	145.0	64.0
Selectivity for Elemental Sulfur, Percent	22.0	75.1	56.3
Selectivity for Sulfur Dioxide, Percent	72.9	23.5	--
Selectivity for Sulfur Trioxide, Percent	5.1	1.4	4.5
Selectivity for Hydrogen Sulfide, Percent	--	--	39.2

Solid samples at various stages of conversion were analyzed by
Mossbauer spectroscopy. The only product identified in this
case was Fe_3O_4. Hence, conversions can be calculated based on
the weight change data. A comparison of conversions calculated
based on the weight change data and those calculated based on
the Mossbauer results is shown in Table IV. These results

Table IV
Results of Iron Oxide-Steam Reaction

Temperature = $695\pm4^{\circ}C$ Steam % = 94.0
Average Rate = 0.33 min^{-1} (Balance N_2)

Run No.	Time Min.	$\dfrac{Fe_3O_4}{FeS_{1.1}}$ wt. ratio By Mossbauer Analysis	Calculated From Mossbauer Results	Calculated From Wt. Loss Data
1	5	0.239	0.210	0.190
2	8	0.254	0.228	0.215
3	10	0.498	0.361	0.362
4	15	-	-	0.440
5	25	-	-	0.890

suggest that reactions (5) and (6) are taking place in this
system. In addition, reaction (6) must be very rapid, so that
at no stage FeO is identified in the solid samples. This is
also found to be consistent with equilibrium calculations.
Studies are now in progress to investigate the role of small
concentrations of air in the system.

   With respect to the overall mechanism, it is worthwhile to
comment briefly on the APCI results of high elemental sulfur
yields, even in the absence of air. The results suggest that the
reaction responsible for the elemental sulfur formation under
these conditions may be reaction (12). The main reason for this
contention is the fact that transition metals and their sulfides
are known to be very good catalysts for H_2S decomposition (11,
12). On the other hand, equilibrium constant for this reaction
at the temperatures of study is rather small. But it is suspected that with the fixed-bed operation and with the possibility of
some sulfur vapor adsorption on the solid, nonequilibrium conditions may be prevailing in the system. As a result, high sulfur
yields could be obtained. This plausible explanation is only
speculative, and more studies are necessary before a definite
conclusion can be drawn. At WVU studies are in progress to
obtain the kinetics of the reactions involved in this scheme.

Analysis of Conversion Data

To correlate the kinetic data obtained from the TGA experiments the grain model, developed by Szekeley and coworkers (13), was used. Accordingly, the solid particle is considered to be composed of small, non-porous grains. The gas has to diffuse through the interstitial spaces and then react at the grain surface. The reaction within the grain is considered to follow the shrinking core model (14). This is represented in Figure 7. In the present system, since the iron oxide (Fe_2O_3) is finely dispersed in the silica matrix, the grain model is considered appropriate.

In the absence of a resistance to the gas-film mass transfer, the overall rate of a noncatalytic gas-solid reaction is influenced by the chemical reaction and by pore diffusion. In general, at low temperatures and with a small particle size, the chemical reaction is likely to control the overall rate. The solid particle in this case reacts uniformly throughout and no concentration gradients exist within the particle. This situation is shown in Figure 8. On the other hand, at high temperatures and with large particles pore diffusion is expected to be important. In this case, reaction occurs at a sharp interface and the process can be described by a shrinking core model as shown in Figure 9. The assumptions used and the mathematical development of this model in the present context are presented in an earlier paper (9). An important assumption made here is that under a given set of conditions the overall reaction rate can be represented by the addition of rates for the pore diffusion and the chemical reaction control regimes, the two asymptotes represented in Figures 8 and 9. The approximate conversion-time equations based on these premises are given as follows (13):

$$t = t_1 F_1 + t_2 F_2 \text{ for a spherical pellet/particle} \quad (14)$$

and

$$t = t_1 F_1 + t_3 F_3 \text{ for a cylindrical pellet/particle} \quad (15)$$

where

$$t_1 = \frac{C_{So}}{bk'(C_{Ao}-C_{Co}/K)}; \quad F_1 = 1-(1-X)^{1/3} \quad (16)$$

$$t_2 = \frac{C_{So} r_o^2 (1+1/K)}{6bD_e(C_{Ao}-C_{Co}/K)}; \quad F_2 = 1-3(1-X)^{2/3} + 2(1-X) \quad (17)$$

$$t_3 = \frac{C_{So} r_o^2 l_o^2 (1+1/K)}{4bD_e(C_{Ao}-C_{Co}/K)(r_o+l_o)^2}; \quad F_3 = X+(1-X)\ln(1-X) \quad (18)$$

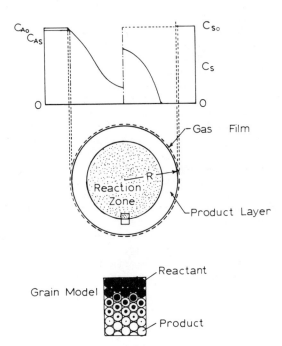

Figure 7.   Schematic diagram of the grain model.

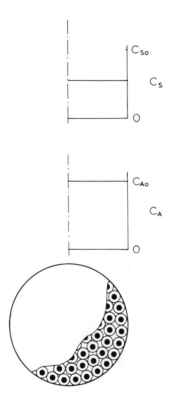

Figure 8. The grain model for a solid undergoing reaction under conditions of chemical reaction control.

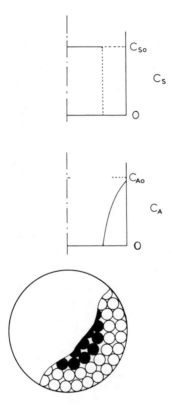

Figure 9. The grain model for a solid undergoing reaction under
conditions of pore diffusion control.

C_{Co} = bulk concentration of the gaseous product, $mole/cm^3$
K   = equilibrium constant
t   = time, sec, X = fractional conversion,
C_{So} = solid reactant concentration, $mole/cm^3$
b   = stoichiometric coefficient
k   = rate constant, sec^{-1}
C_{Ao} = bulk concentration of the gaseous reactant, $mole/cm^3$
r_0 = particle radius, cm
l_0 = length (for a cylinder), cm
D_e = effective diffusivity, cm^2/sec

Using the above equations and the experimental results in
the asymptotic regions, intrinsic rate constants (k) and effective diffusivity (D_e) values were determined for reactions (1)
to (4). These are given in Table V. Similar studies are in
progress for the steam-air regeneration system. Notably, the
activation energy value obtained for the intrinsic sulfidation
reaction is unusually low. This is attributed to the reaction
mechanism. The rate-controlling step seems to be the dissocia-
tive adsorption of H_2S on the solid surface, which has a low
activation energy. Similar conclusions have been drawn by
Turkdogan and Worrell (15,16) in their studies on the sulfidation
of iron strips. This has a significance in the steam regeneration system, wherein the elemental sulfur formation is speculated
to be due to a catalytic decomposition of H_2S.
     To validate the grain model used and to substantiate the
accuracy of the experimental data, the k and D_e values given
in Table V were used to predict weight change curves for the
reactions under different conditions. In all the cases the
predicted and the experimental curves matched very well. Two
such typical comparisons are illustrated in Figures 10 and 11
(9,10). Figure 10 shows the weight change curve for the reaction
of iron oxide with a simulated coal gas, which was calculated
by a combination of results for the reduction and the sulfidation
reactions. In Figure 11 results are shown for the reaction of
iron sulfide with SO_2, which illustrate a close agreement between the calculated values and the experimental values obtained
by different analytical techniques.

Conclusion

     The iron oxide-silica sorbent with 45 weight percent
Fe_2O_3 appears to be very effective and efficient as a sorbent
for the bulk sulfur removal from a hot low-/medium-Btu gas.
The active sorbent form, which reacts with H_2S, is the metallic
iron, Fe or the oxide FeO. This is due to the initial rapid
reduction of Fe_2O_3 by the H_2 and the CO present in the fuel
gas. On sulfidation of the sorbent a nonstoichiometric
pyrrhotite, $FeS_{1.1}$, is formed. This reaction of Fe with H_2S is
much slower compared to the reduction reaction.

Table V

Stoichiometric Coefficient, Rate Constant and Effective Diffusivity Values
Obtained for Different Reactions Using the Grain Model.

Reaction

(1) $Fe_2O_3 + 3/2 (H_2+CO) \quad 2 Fe + 3/2 (H_2O+CO_2)$

$b = 2/3$

$k'_r = 35550 \exp (-14,800/RT) \text{ sec}^{-1}$

$D_{er} = 0.06462 \exp (-5,000/RT) \text{ cm}^2/\text{sec}$

(2) $Fe + 1.1 H_2S \quad FeS_{1.1} + 1.1 H_2$

$b = 0.91$

$k'_s = 158 \exp (-3,300/RT) \text{ sec}^{-1}$

$D_{es} = 0.24 \exp (-4,300/RT) \text{ cm}^2/\text{sec}$

(3) $3 FeS_{1.1} + 2 SO_2 \quad Fe_3O_4 + 1.55 S_2$

$b = 3/2$

$K'_{SO_2} = 7.8916 \times 10^3 \exp (-17,500/RT) \text{ sec}^{-1}$

$D_{e(SO_2)} = 0.246 \exp (-6,140/RT) \text{ cm}^2/\text{sec}$

(4) $2 FeS_{1.1} + 3.7 O_2 \quad Fe_2O_3 + 2.2 SO_2$

$b = 0.54$

$k'_{O_2} = 1 \times 10^5 \exp (-15,630/RT) \text{ sec}^{-1}$

$D_{E(O_2)} = 9.1 \exp (-5,370/RT) \text{ cm}^2/\text{sec}$

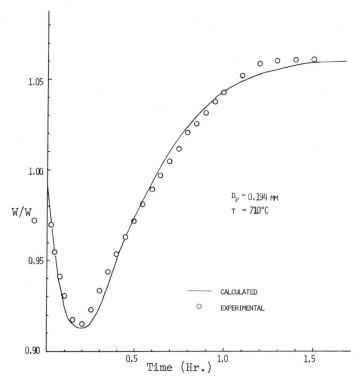

Figure 10.   Comparison of calculated weight change curve with the experimental data for simultaneous reduction-sulfidation.

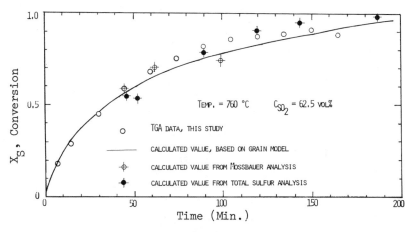

Figure 11.   Comparison of the experimental data with the calculated values for pellets with size of 1.5 x 6.0 mm.

The key factor in the development of a commerical hot-gas desulfurization process is the regeneration of the spent sorbent. Two methods are discussed here: regeneration with a air-SO_2 mixture and regeneration with a steam-air mixture. The reaction of iron sulfide with air is very fast and exothermic, and forms Fe_2O_3 and SO_2 as the only products. Consequently the reaction rate is largely controlled by the pore diffusion mechanism. Reactions of iron sulfide with SO_2 and with steam are relatively much slower, and form Fe_3O_4 and elemental sulfur. The overall rate in these cases is largely controlled by the intrinsic chemical reaction.

Acknowledgments

The financial assistance for this work is provided by the U.S. Department of Energy, Morgantown Energy Technology Center.

Literature Cited

1. MERC Hot Gas Cleanup Task Force, "Chemistry of Hot Gas Cleanup in Coal Gasification and Combustion," MERC/SP-78/2, February 1978.
2. Lewis, P. S.; Shultz, F. G.; Wallace, W. E. Jr., "Sulfur Removal from Hot Producer Gas," presented at the 166th National Meeting, ACS Div. Fuel Chem., Chicago, Aug. 26-31, 1973.
3. Jenkins, D. M.; Lemmon, A. W. Jr., "Supplemental Report on Evaluation of High-Temperature Desulfurization Systems for Low- and Intermediate-Btu Gas," Battelle Columbus Labs. for the Tennessee Valley Authority Power Research Staff, Dec. 31, 1975.
4. Oldaker, E. C.; Gillmore, D. W., 172nd National Meeting, ACS, Div. Fuel Chem. Preprints 1976, 21(4), 79.
5. Grindley, T.; Steinfeld, G., "Development and Testing of Regenerable Hot Coal Gas Desulfurization Sorbents," Report DOE/MC/16545-1125 (DE82011114), Oct. 1981
6. Karr, C.; Rahfuse, R. V. Jr.; Langdon, P. E., J. Appl. Chem. Biotechnol 1972, 22, 613.
7. Schrodt, J. T.; Best, J. E., A.I.Ch.E. Symp. Ser. No. 175 1978, 74, 184.
8. Joshi, D.; Olson, J. H.; Hayes, M. L.; Shah, V., "Hot, Low-Btu Producer Gas Desulfurization in Fixed-Bed of Iron Oxide-Fly Ash," USDOE Contract FE-77-3-01-2757, Final Report, June 1, 1979.
9. Tamhankar, S. S.; Hasatani, M.; Wen, C. Y., Chem. Engng. Sci. 1981, 36, 1181.
10. Tseng, S. C.; Tamhankar, S. S.; Wen, C. Y., Chem. Engng. Sci. 1981, 36, 1287.

11. Fukuda, K.; Dokiya, M.; Kameyama, T.; Kotera, Y., Ind. Eng. Chem. Fundam. 1978, 17, 243.
12. Worrell, W. L.; Kaplan, H. I., in Heterogeneous Kinetics at Elevated Temperature, Plenum Press, 1979, p. 113.
13. Szekeley, J.; Evans, J. W.; Sohn, H. Y., "Gas-Solid Reactions," Academic Press, New York, 1976
14. Levenspiel, O., "Chemical Reaction Engineering," 2nd Edn., Wiley, New York 1972, p. 372.
15. Turkdogan, E. T., Trans. Met. Soc. AIME 1968, 242, 1655.
16. Worrell, W. L.; Turkdogan, E. T., Trans. Met. Soc. AIME 1968, 242, 1673.

RECEIVED December 7, 1982

INDEX

INDEX

A

Absorber
coal gasification methane recovery,
dimensions and
characteristics 241, 242t
H₂S sorption by iron oxide
reaction 265
Absorption
carbon dioxide, after sulfur
removal 40, 42–44f
carbonyl sulfide by liquid CO_2 ...33, 35, 36f
properties of physical absorbents 37t
slurry, carbon dioxide removal 42, 43f
and stripping process,
methane recovery,
description 236, 240f, 241–45
sulfur by liquid CO_2 ....33, 34f, 35
TGA studies of H₂S reactions of
iron oxide sorbent 262, 264f
Acetone, effective molar or partial
molar volume of CO_2 .... 55t
Acid gas removal—*See* AGR
Activation energy, diffusion 92
Adsorbate loading, average, equation 174
Adsorbent
dimensionless temperature
profiles 185–92
dimensionless temperature, equation for non-isothermal LDF
model 174, 176
paraffin separation 159
PSA process, carbon-based
molecular sieve 156
regeneration time effect on product
quality 230, 231f
suitable, characteristics 146
temperature profile, complete heat
transfer control 177
Adsorbent mass, gas sorption
kinetics 173
Adsorption bed, two component
mixture, PSA cell model 197
Adsorption capacities, nitrogen clinoptilolite adsorption 226, 228f, 229f
Adsorption
gas
criteria for when to use 146
process state of the art 145–69
purification process 162
various commercial processes ..147–67
gas separation, future
directions 164, 166–67

Adsorption—*Continued*
isotherm, N₂ on clinoptilolite 218f
isothermal, kinetics equation 177
relative, methane and nitrogen clinoptilolite adsorption ..215, 220, 221t
selective, methane and nitrogen
zeolite separation 213–32
Adsorption bed, two component mixture, PSA cell model 197
Adsorption capacity, nitrogen clinoptilolite adsorption 226, 228f, 229f
Adsorptivity, effect of pressure 154
Advantages, PSA processes 152
AGR
Claus plant process 29
CNG process description 30–35
CO_2 condensation 33–35
coal gasification, problems 29
discussion and applications 27–28
membrane separation from methane 135
process 28, 30
Air
oxidation regeneration of sulfided
sorbent 261
oxygen enrichment by membrane
process 142
Argon, apparent heats of adsorption
over various clinoptilolites 216t
Arrhenius rate expression, diffusion
coefficient 93
Assink, NMR studies of sorption and
transport in glassy polymers 106
Axial pressure profile, PSPP process .. 156

B

Bead activated carbon, use in solvent
vapor removal 164
Benzene
effective molar or partial molar
volume of CO_2 .... 55t
methane solubility at various
pressures 236, 237t
Bergbau Forschung nitrogen production process 158f
BET surface area, correlation with ion
exchange degree in clinoptilolite 216t
Biot number
dimensionless uptake equation 176
effect on dimensionless temperature
profiles 191f, 192
effect on uptake curves 185, 190f

Blowdown equations, PSA cell
    model 200, 201
Boundary conditions, kinetics, LDF
    model 174
Butane
    hydrogen recovery from isobutane
        production 140, 141*f*
    methane solubility at various
        pressures 236, 237*t*

C

Calcium
    effect on clinoptilolite nitrogen
        retention 223*f*, 224*f*
    regeneration product in H₂S
        removal 360*t*
Calibration, problems in gas trans-
    mission studies 77, 78
Carbon-13 NMR
    PVC-TCP systems 95*f*
    relaxation rates 94, 96*t*
Carbon, activated bead, solvent
    vapor removal 164
Carbon dioxide
    cellulose acetate permeation rates
        and separation factors 134, 135*t*
    condensation in AGR process 33–35
    in conditioned polycarbonate 120*f*
    crystallization in CNG process ... 38–40
    dew point in water-free crude gas ..33, 34*f*
    effect of mixed gas on
        permeability 62, 63*f*
    effective molar or partial molar
        volume in various
        environments 55*t*
    liquid, physical properties 35–36*f*
    permeability and time lag data for
        gas transmission study 80, 81*t*
    phase diagram 32*f*
    in polycarbonate, sorption
        isotherm 117, 118*f*
    pressure effect on relaxation rate
        and apparent diffusion coeffi-
        cient in PVC 104, 105*f*
    PVC sorption ranges 103
    recovery use in enhanced oil
        recovery 136
    regeneration by triple point
        crystallization 35–41*f*
    removal
        from natural gas 135
        after sulfurous compound
            absorption 40, 42–44*f*
        slurry absorption 42, 43*f*
    solid, solubility in various solvents 42, 43*f*
    sorption and transport parameters
        in lexan polycarbonate 65*f*
    time lag diffusion in polycarbonate 122*f*
    transport process in SRM 1470 83
    triple point crystallization 39*f*

Carbon disulfide, recovery, fluidized-
    bed process 164
Carbon monoxide
    cellulose acetate permeation rates
        and separation factors 134, 135*t*
    effect of concentration of TCP
        in PVC 98*f*
    PVC-TCP gas transport and C-13
        NMR parameters 96*t*
Carbon tetrachloride, effective molar
    or partial molar volume of CO₂ .. 55*t*
Carbonyl sulfide, absorption by
    liquid CO₂ 33, 35, 36*f*
Cation deficiency, effect in naturally
    occurring clinoptilolite 214
Cell model, PSA light gas
    separation 197–210
Cellulose acetate membrane
    cross section 127*f*
    data 134, 135*t*
    discussion 126, 127*f*
    effect of hydrocarbons 131
    hydrogen recovery 140–42*t*
    water vapor removal 139
Chain separation, activation energy
    of diffusion 92
Chain stiffness 92
Chemical reaction, effect on adsorp-
    tion rate in grain model 271
Chemisorption, AGR process 28
Chlorobenzene, effective molar or par-
    tial molar volume of CO₂ 55*t*
Chromatography, commercial gas
    separation process 162, 163*f*
Claus process
    AGR process 29
    desulfurization process conversion
        to elemental sulfur 261
    reactions 268
Clinoptilolite
    See also Zeolite
    characterization 214
    ion exchange, degree correlated
        with surface area and separa-
        tion ability 216*t*
Closed cycle unit, triple point
    crystallization 38, 39*f*
CNG process
    CO₂ crystallization 38–40
    description and features 30–45
    final CO₂ removal 40, 42–44*f*
CNG Research Company, proprietary
    AGR process development 28
Coal gasification
    AGR process development goals 30
    gas composition 29
    major contaminant 255
    membrane hydrogen recovery 140
    products 235
    system schematic 256*f*

Cobalt, regeneration product in
 H₂S removal 360t
Concentration dependence
 local diffusion coefficient 57
 polymer properties, matrix model .. 113
 sorption and transport model 99
Concentration dependence model,
 sorption and transport in glassy
 polymer system 112
Concentration polarization, erratic
 permeability coefficient 14
Concentration profile, triple-point
 crystallizer 40, 41f
Condensation, CO₂ in AGR process ..38–40
Conditioning, penetrant induced in
 glassy polymers 48–57
Contaminants, trace, coal gasification 29
Conversion time, equations for
 grain model 271
Conversions, H₂S, various sorbents 256
Cooperative main chain motions
 effect on diffusion coefficient of
 glassy polymers 114
 in PVC-CO₂ 101, 104
Copper, regeneration product in
 H₂S removal 360t
Cost, criteria used in H₂S sorbent
 selection................................. 256
Crystallization
 CO₂ in CNG process 38–40
 triple point, CO₂ 35, 41f
Cycles, gas adsorption process 147–52
Cyclohexane, methane solubility at
 various pressures 236, 237t

D

Density, adsorbent mass, LDF model 173
Dew point, CO₂ in coal gasification .. 33, 34f
Diffuse movements, polymer penetrant
 molecules 60
Diffusion
 See also Transport
 clinoptilolite adsorption of mixed
 gas systems 220, 222–24t
 effect of cooperative main-chain
 motions 97, 104
 mechanism in polymers 93–104
 mixed gas, zeolite 222, 224f
 molecular model of gases in a
 polymeric matrix 91–93
 rate determining step 92
Diffusion coefficient
 apparent
 effect of CO₂ pressure in
 PVC 104, 105f
 effect of concentration of
 TCP in PVC 98f
 H₂ and CO 94, 96t
 Arrhenius form 91
 clinoptilolite nitrogen and methane
 adsorption 219f, 220, 221t

Diffusion coefficient—*Continued*
 concentration dependency 59
 effect of main chain molecular
 motions 104
 equation 93
 glassy polymers 114, 115
 methane and nitrogen, measure-
 ment apparatus 217f
 simplified expressions 115, 116
Diffusivity
 See also Permeability
 effective, various grain model
 reactions 276t
 gas in glassy polymers 115
 intraparticle, use in nitrogen
 production 156
Dimensionless time, pressurization
 step in PSA cell model 200
Dimethyl ether, physical properties 37t
Disadvantages, PSA processes 152
Displacement purge cycle
 discussion 159–62, 166
 gas adsorption processes 149, 150f
Distillation vs. gas adsorption,
 when to use 146
Dual mode model
 discussion, Raucher and Sefcik
 and Assink NMR study 66
 permeability pressure
 dependence 119, 120f
 sorption and transport,
 discussion 100, 101, 106–8

E

Efficiency, criteria used in selecting
 sorbent for H₂S removal 256
Elf Aquitaine chromatographic
 separator 163f
Enhanced oil recovery, application
 for membrane systems 136
Energy calculations, various, coal
 gasification absorption and stripping methane recovery process 245–47
Energy efficiency, physical absorption
 AGR processes 29
Energy requirement, absorption and
 stripping methane recovery
 processes 243, 245t
Enrichment
 helium, various feed gas
 compositions 14, 16f–20f
 light component in PSA single
 column process 202
Enrichment permeabilities and separation factors, helium 14–21
Equations
 permeability of membrane 6, 8, 9t
 various, PSA cell model 199–204
Equilibrium parameter
 dimensionless uptake equation 176

Equilibrium parameter—*Continued*
  effect on dimensionless tempera-
    ture profiles 185, 189*f*
  effect on the uptake curves 185, 188*f*
Ethane, cellulose acetate permeation
    rates and separation factors ..134, 135*t*
Ethanol, methane solubility at
    various pressures 236, 237*t*
Ethylene, effect on CO_2 sorption 62

F

Feed composition, effect on enrich-
    ment in two column PSA process 208
Fickian diffusion model, sorption
    mass transfer 172
Fick's law
  concentration dependent equation .. 99
  diffusion equation 131
  dual mobility transport model 57
  transport of gases in polymers 90
Flow rate
  effect on membrane gas
    enrichment 136, 138*f*
  various component gases in absorp-
    tion and stripping methane
    recovery process 244*t*
Fluidized-bed/moving-bed process,
    solvent vapor removal 164–66
Flux
  characterization of SRM 76
  effect of mixed gas 48
  glassy polymer 57
  membrane separations 14, 131
  reductions by non interacting
    penetrants 62
  steady state, equation 90
  time dependent decline 138*f*, 139
Fourier conduction, adsorbent heat
    transfer 172
Frequency motions, high and low,
    polymer diffusion 92
Fuel, synthetic, AGR step in produc-
    tion from coal 28

G

Gas
  adsorption process
    description 145, 147, 148*t*
  bulk separation
    process ..148*t*, 152–62, 164, 166–67
  coal gasification, TGA study of iron
    sorbent reactions 262
  composition effect on separation
    efficiency 132–34
  control of solubility coefficient in
    polymer interactions 113
  crude coal gasifier 28
  deep sea diving, feed gas
    composition 6
  dehydration by membranes 139

Gas—*Continued*
  enrichment and recovery in single
    column PSA process 205, 206*f*
  enrichment and recovery in two
    column PSA process 205, 207*f*
  feed, composition for helium
    recovery 6*t*
  light, PSA recovery and
    purification 195–210
  mixed
    dual sorption and transport
      model 61–64
    effect on permeability 48
  non-isothermal sorption kinetics ..171–94
  purification 148*t*, 162
  residual and permeate stream
    compositions related to stage
    cut 136–38*f*
  separation
    basic equations 131, 132
    mechanism 126
    parameters affecting
      efficiency 132–34
    technical breakthrough in the
      application of membranes .. 126
  sorbed molecules, effect on polymer
    equilibrium 103
  sorption by glassy polymers,
    molecular environments 52
  Standard Reference Materials for
    transmission measurements 75–88
  synthesis, solubility of solid CO_2 ..42, 43*f*
  transport and C-13 NMR
    parameters, PVC-TCP 96*t*
  transport mechanism in glassy
    polymers 89
  treatment in CNG AGR process 31*f*
  two component mixture, PSA
    cell model 197
Gasification, AGR step in synthetic
    fuels production 28
Gas oil, methane solubility at
    various pressures 236, 237*t*
Gas/polymer matrix model
  See also Matrix model
  gas sorption and transport
    polymers 112–16
Grain model
  chemical reaction control conditions 273*f*
  correlation of kinetic data for H_2S
    iron oxide sorption 271
  pore diffusion control conditions 274*f*
  schematic diagram 272*f*
  validity 275

H

Heat of adsorption
  measurements for clinoptilolite
    methane and nitrogen
    separation 214, 215, 216*t*
  n-pentane uptake on zeolite 180

Heat capacity
adsorbent mass, LDF model 173
liquid-solid slurry absorbent 40
Helium
cellulose acetate permeation rates
and separation factors 134, 135t
enrichment
permeabilities and separation
factors 14–21
various feed gas
compositions 14, 16f–20f
and methane separation, effect of
pressure on single column
PSA 208t
recovery
experimental loop schematic 2, 4f, 5f
as function of pressure drop 14, 15f
hypothetical system 21–23f
membranes 1–24
purity goals and requirements 3
Henry's law
concentration dependent equation .. 99
diffusion equation 131
glass polymer conditioning 52
polymer penetrant mode 60
transport of gases in polymers 90
Heat transfer coefficient, external,
zeolite LDF model 184
Heat transfer resistance, effect on
uptake curves 180
Hexane, methane solubility at
various pressures 236, 237t
High pressure channel spacer, spiral-
wound elements 128
Hollow fiber, configuration of
membrane elements 128
Hydrocarbons
effect on cellulose acetate
membranes 131
gas stream dehydration by cellulose
acetate membranes 139
Hydrogen
cellulose acetate permeation rates
and separation factors 134, 135t
effect of concentration of TCP
in PVC 98f
PVC-TCP gas transport and C-13
NMR parameters 96t
recovery 140–42t, 166
Hydrogen sulfide
cellulose acetate permeation rates
and separation factors 134, 135t
removal from coal gasification
products 255–78
removal from natural gas 135

I

Inert purge cycle,
discussion 147, 159–62, 166
Inhomogeneity, obstacle to the development of better SRM's 85
Integrity verification,
membranes 11, 12f, 13t
Interchain cohesion 92
Intersegmental gaps, polymers, redistribution by high pressure
penetrants 52
Ion exchange, clinoptilolite, degree
correlated with surface area and
separation ability 216t
Iron, regeneration product in
H₂S removal 360t
Iron oxide
results of steam regeneration
reaction 270
sorbent for high temperature
H₂S removal 255–78
sulfided, effect of temperature on
weight loss in SO₂ atmospheres 266f
weight change curve of reaction
with simulated coal gas 276f
Iron sulfide, reaction with SO₂ 276f
Isooctane, zeolite uptake data, LDF
kinetics model 178, 192
Isomer separation processes,
discussion 159–63f
Isomerization process, total, paraffin
separation 159, 161f
IsoSiv process, paraffin separation 159
Isosteric enthalpy of sorption, effect of
sorbed gas concentration 114
Isotherm
adsorption of N₂ on clinoptilolite .. 218f
equations for PSA cell model 199
sorption, CO₂ in polycarbonate ..117, 118f
sorption, matrix vs. dual mode
model for glassy polymers 118f
Isothermal model, estimation of
diffusivity dependence 185

K

Kinetics
adsorption, non-isothermal LDF
model equations 173–78
criteria used in H₂S sorbent
selection 256
high temperature H₂S sorption by
iron oxides 262–70
high temperature H₂S sorption by
selected metal oxides 256, 258f
mathematical models, non-isothermal sorption 172
methane and nitrogen clinoptilolite
adsorption 215, 220
Knudsen effect, effective thermal
conductivity 180

L

Langmuir, polymer penetrant mode .. 60
Langmuir isotherm, glassy polymer
   conditioning 52
Langmuir sorption capacity
   CO_2 in various polymers, experi-
     mental compared with
     calculated 58f
   glassy polymers 53, 54f
Langmuir system, equilibrium
   parameter 185
LDF model, non-isothermal 173
Lewis number, modified, reciprocal
   dimensionless uptake equation 176
   effect on dimensionless temperature
     profiles 185, 187f
   effect on uptake curves 185, 186f
Linear driving force model 173

M

Main chain motions
   C-13 rotating-frame relaxation
     rates 94
   effect on diffusion of gases 97
   of PVC, effect of CO_2 101
Magnetite, product of pyrrhotite and
   SO_2 in iron oxide regeneration .. 265
Manganese, regeneration product in
   H_2S removal 360t
Manometers, electronic measuring
   technique to monitor gas flux 76
Mass balance, methane and nitrogen
   PSA clinoptilolite process ... 226, 227f
Material balance, equations, PSA
   cell model 199, 204
Mathematical model, behavior of
   adsorbent columns 173
Matrix model
   See also Gas polymer matrix model
   permeability pressure
     dependence 119, 120f
Mechanism
   glassy polymer diffusion 101–4, 112
   high temperature H_2S sorption
     by iron oxides 262–70
   in polymer diffusion 93–101
Mean free paths, isooctane and
   n-pentane, LDF model 184
Membranes
   cellulose acetate
     cross section of asymmetric 127f
     effect of water 21
     flat-sheet preparation 126
     permeation and selectivity data .. 134
     spiralwound semipermeable 6, 7f
     definition of permeability 6
     effect of high pressure drop 14
     gas separations for chemical proc-
       esses and energy applications 125–42

Membranes—*Continued*
   integrity verification 11, 12f, 13t
   parameters affecting gas separation
     efficiency 132–34
   semipermeable, helium recovery 1–24
   spiralwound, discussion 128–31
   various applications for gas
     separations 135–42t
   various element configurations ...128–31
Methane
   apparent heats of adsorption over
     various clinoptilolites 216t
   cellulose acetate permeation rates
     and separation factors 134, 135t
   diffusion coefficient in clinoptilolite
     adsorption 219f, 221
   effect on clinoptilolite nitrogen
     adsorption 222, 224t
   and helium separation effect of
     pressure on single column
     PSA 208t
   membrane AGR 135
   separation from H_2 and CO by
     adsorption and stripping
     process 235–46
   variables for PSA on
     clinoptilolite ...226, 227t, 228f, 229f
   zeolite adsorption 213–32
Methanol
   methane solubility at various
     pressures 236, 237t
   physical properties 35–37t
Methyl acetate, effective molar of partial molar volume of CO_2 .... 55t
Model
   cell, PSA light gas separation ...197–210
   dual mobility transport 57
   glassy polymers, matrix vs.
     dual mode 118f, 120f, 122f
   non-isothermal sorption kinetics 172
Mole fraction, various component
   gases in absorption and stripping
   methane recovery process 244t
Mole fraction profile, PSA cell
   model 200–203
Molecular environments, gas sorption
   by glassy polymers 52
Molecular sieve
   carbon based, PSA process
     adsorbent 156
   paraffin separation adsorbent 159
Molecular theory, diffusion of gases
   in a polymeric matrix 91
Molybdenum, regeneration product in
   H_2S removal 360t
Mössbauer analysis, sulfided iron
   oxide regeneration 265, 267t
Moving bed processes, future growth 166
Multiple bed process, PSPP 154

N

Naphtha, heavy, methane solubility at various pressures 236, 237t
Nickel, regeneration product in H_2S removal 360t
Nitrogen
  apparent heats of adsorption over various clinoptilolites 216t
  cellulose acetate permeation rates and separation factors ...134, 135t
  diffusion coefficient in clinoptilolite adsorption 219f, 221
  effect of methane on clinoptilolite adsorption 222, 224t
  permeability and time lag data for gas transmission study 80, 81t
  permeation rate related to membrane integrity 11, 13t
  production by PSA process 156, 158f–60f, 164
  related to helium enrichment 16f, 18f, 19f
  variables for PSA on clinoptilolite ...226, 227t, 228f, 229f
  zeolite adsorption 213–32
NMR, Assink study, support of dual mode model 66

O

Octane, methane solubility at various pressures 236, 237t
OlefinSiv, butene separation process .. 162
Oxygen
  enrichment of air by membrane process 142
  permeability and time lag data for gas transmission study 80, 81t
  recovery by PSA, future 164
  related to helium enrichment 16f, 18f, 19f

P

Paraffins, separation by displacement purge and inert purge cycles ..159, 161f
Parameters, operating, typical spiral-wound membrane units 131
Parametric pumping 154
Partial immobilization model, gas transmission study 83
Penetrant
  diameter effect on activation energy of diffusion 92
  effect on polymer cooperative motions 103
  glassy polymers, effective molecular volume 53, 54f
  induced conditioning in glassy polymers 48–57
Pentane, methane solubility at various pressures 236, 237t

n-Pentane, zeolite uptake data, LDF kinetics model 178, 192
Permeability
  See also Diffusivity
  calculated, matrix vs. dual mode models 119, 120f
  coefficients, helium 14, 21
  comparative data 13t
  data, H_2 and CO in PVC-TCP 96t
  effect of various mixed gas concentrations 62, 63f
  glassy polymer, various partial pressures 48, 49f
  membrane, equations 6, 8, 9t
  SMR study, various gases 82f
Permeability coefficient, steady state equation 90
Permeability data, SRM evaluation study 80, 81t
Permeance—See Permeability
Permeate channel spacer, spiralwound elements 128
Permeate enrichment, effect of feed composition 133f
Permeate tube, spiralwound elements 128
Permeation
  See also Transport
  mixed gas, dual mode sorption and transport models 47–71
Permeation rates, various gases in cellulose acetate membranes 134
Permselectivity, membranes, discussion 8, 11t, 19f, 20f, 22f
PET
  CO_2, Langmuir sorption capacity .. 56f
  gas transmission study, various gases 77
  permeabilities and related quantities of various materials 84t
  SRM inhomogeneity 83
Phase diagram, CO_2 .... 32f
Physical absorption, AGR process 28, 29
Physical model, glass polymers conditioning 52
Physical strength, criteria used in selecting H_2S sorbent 256
Plasticization
  effect on permselectivity of membranes 48
  rubber polymers vs. glassy polymers 59
Plate and frame, configurations of membrane elements 128
Poly(acrylonitrile), CO_2 Langmuir sorption capacity 56f
POLYBED PSA process 152, 153f
Poly(benzyl methacrylate), CO_2 Langmuir sorption capacity 56f
Polycarbonate
  CO_2 Langmuir sorption capacity 56f

Polycarbonate—*Continued*
  CO_2 sorption isotherm 117, 118*f*
  CO_2 in conditioned 120*f*
Poly(ethyl methacrylate), CO_2 Lang-
  muir sorption capacity 56*f*
Poly(ethylene terephthalate)—*See*
  PET
Polymer
  conditioning 103
  effect of high gas conditioning
    pressures 52
  glassy
    concentration dependent sorption
      and transport models 104–9
    conditioned nonequilibrium 52
    diffusion coefficient equation 115
    dual mobility transport model .. 57
    equilibrium densified 52
    gas transport and cooperative
      main chain motions 89–110
    Langmuir sorption capacity 53, 54*f*
    mathematical expression of
      sorption 117
    matrix and dual mode models as
      physical models 121
    matrix sorption and transport
      model 111–24
    penetrant induced conditioning .. 48
    segmental motion 49, 51*f*
    sorption and transport 97–109, 115
    various, CO_2 Langmuir sorption
      capacity 56*f*
  high and low frequency diffusion
    motions 92
  mathematical expression of gas
    transport 117, 119–21
Polymer chains 91, 92
Poly(methyl methacrylate)
  CO_2 Langmuir sorption capacity 56*f*
  effect of ethylene on CO_2 sorption
    level 63*f*
Poly(phenyl methacrylate, CO_2
  Langmuir sorption capacity 56*f*
Poly(terephthalate), local effective diffusion coefficient for CO_2 .... 58*f*
Polyvinyl chloride, CO_2 diffusion
  coefficient 59
Pore accessibility,
  clinoptilolite 214, 215, 217*f*
Pore diffusion, effect on rate in grain
  model 271
Pressure
  effect on adsorptivity 154
  effect on permeability matrix vs.
    dual mode model 119, 120*f*
  effect on single column process for
    helium and methane
    separation 208*t*
  effect on time lag, matrix vs.
    dual model 121, 122*f*
  inside adsorbent mass, LDF model 173

Pressure—*Continued*
  various component gases in adsorp-
    tion and stripping methane
    recovery process 244*t*
Pressure differential, factor in mem-
  brane gas separation 132
Pressure profiles, PSPP process 156, 157*f*
Pressure ratio, effect on separation
  efficiency 132–34
Pressure sensors, calibration
  instability 78
Pressure step, constant, equations for
  PSA cell model 201
Pressure swing—*See also* PSA
Pressure swing adsorption, recovery
  and purification of light gases ..195–211
Pressure swing cycle, gas adsorption
  process 149, 150*f*
Pressure swing parametric
  pumping 152, 154–56
Pressurization, equations, PSA cell
  model 200, 201
Propane
  loss in absorption and stripping
    methane recovery ...236, 238*f*, 238*t*
  methane solubility at various
    pressures 236, 237*t*
PSA—*See also* Pressure swing
PSA process
  basic comparison to Polybed PSA
    and PSPP 157*t*
  discussion 152–59, 164, 166
  methane and nitrogen clinoptilolite
    adsorption ...222, 225*f*, 226–30
  single column bulk
    separation 196*f*, 202, 203
  two column bulk separation 198*f*, 203, 204
Pumps, parametric, one-bed and
  three-bed pressure swing 155*f*
PURASIV HR process, solvent
  vapor removal 165*f*
Purification, gas 148*t*
PVC
  See also Poly(vinylchloride)
  antiplasticization-plasticization 93
PVC-CO_2 system, cooperative main-
  chain motions 101, 104
PVC-TCP system, gas transport and
  C-13 NMR parameters 96*t*
Pyrrhotite, product of iron reaction
  with H_2S 265

R

Rate constant, various grain model
  reactions 276*t*
Reaction rate, representation for
  grain model 271
Regeneration
  adsorber, PSA clinoptilolite
    process 230, 231*f*

Regeneration—*Continued*
criteria used in selecting H₂S
sorbent 256
in air, various H₂S sorbents 360*t*
effect of water content on product
distribution 269*t*
options, hot gas desulfurization
process 261
reactions
iron oxide by SO₂ recycle
method 265–68
sulfided iron oxide 268–70
Relative humidity, gas, effect on
separation process 134
Relaxation rate
C-13 NMR, PVC-CO₂ system 102*t*
effect of CO₂ pressure in PVC ..104, 105*f*
effect of concentration of TCP
in PVC 98*f*
PVC-CO₂ system 101, 102*t*

S

Selectivity, effective, effect of pressure
ratio and membrane separation
factor 133*f*
Separation ability, correlation with ion
exchange degree in clinoptilolite 216*t*
Separation factor
See also Permselectivity
effect on separation efficiency 132–34
methane and nitrogen clinoptilolite
PSA process 226, 228*f*, 229*f*
various gases in cellulose acetate
membranes 134
Separation
bulk gas processes 148*t*, 152
bulk gas, PSA cell model 202, 203
normal and iso-paraffins 159
SEPAREX—*See* Cellulose acetate
membranes
Sewage wastewater treatment, membrane oxygenation systems 142
Single bed process, PSPP 154
Single column recovery process, PSA
cell model 195
Solubility
methane, various solvents and
pressures 236, 237*t*
solid CO₂ in several solvents 42, 43*f*
Solubility coefficient, gases in glassy
polymers, equation 114, 116
Solvent vapors, removal from air
streams 164, 165*f*
Sorbate mass transfer, LDF model 172
Sorbent selection, high temperature
H₂S removal from coal gasification products 256
Sorption
See also Uptake
adsorbent non-isothermality
kinetics 172

Sorption—*Continued*
concentration dependence in glassy
polymers 116
concentration dependent model 104–9
concept of unrelaxed volume 48–61
dual mode vs. concentration
dependent models 112
dual mode model ..48, 51*f*, 52, 60, 106–8
equations 90
glassy polymers 111, 117
isooctane on zeolite 179*f*, 181*f*
mixed gas model 61–64
n-pentane on zeolite 182*f*, 183*f*
zero concentration, gas glassy
polymer systems 115
Sorption data, matrix model
analysis 116–21
Sorption model, various, discussion ..99–101
Sorption parameters, CO₂ in lexan
polycarbonate 65*t*
Spiralwound membrane element
configuration 128–31
SRM 1470—*See* PET
SRM, gas transmission study,
problems 85
Stage cut, membrane gas separation,
definition 136, 137*f*
Standard Reference Material,
definition 75
Steam-air mixture, sulfided iron oxide
regeneration 268–70
Stoichiometric coefficient, various
grain model reactions 276*t*
Strippers, coal gasification methane
recovery, dimensions and
characteristics 241, 242*t*
Sulfur capture capacity, criteria used
in selecting H₂S sorbent 256
Sulfur dioxide, sulfided iron oxide
regeneration 265–68
Sulfur, absorption by liquid
CO₂ 33, 34*f*, 35

T

Temperature
gas, effect on the separation process 134
glass transition, depression by
diluents 114
various component gases, absorption and stripping methane
recovery process 244*t*
Temperature profiles
adsorbent, dimensionless 185–92
isooctane uptake on zeolite 180
n-pentane uptake on zeolite 180
Temperature swing, gas processes ..147, 162
TGA curves, reactions of iron oxide
in different gas atmospheres 265*f*
Thermal conductivity, effective
adsorbent mass, LDF model 173
zeolites, LDF model 180

Thermogravimetric analyzer (TGA),
  use in study of reaction mechanisms and kinetics of H_2S
  sorption 262
Time, dimensionless, various stages
  of PSA cell model 200, 201
Time lag
  diffusion of CO_2 in polycarbonate .. 122f
  diffusion of a gas across a planar
    membrane 119, 121–22f
  technique for measuring permea-
    tion gas 77
  SMR study, various gases 80–82f
Toray reverse PSA nitrogen
  production process 160f
Transient transport measurements,
  validity of sorption and transport
  model 119, 121–22f
Transmission
  See also Permeation
  calibration of SRM measurement
    system 77–79
  rates in various packaging materials   76
  study experimental design   79
Transport
  See also Diffusion
  See also Permeation
  concentration dependent
    model 104–9, 116
  dual mode vs. concentration
    dependent models   112
  equation, dual mode model   60
  equations   90
  glassy polymers 111, 117, 119–21
  gas in PVC-TCP system 94–97
  inconsistencies in dual mode
    model 106–8
  mixed gas model 61–64
  process, CO_2 in SRM 1470   83
  transient state 119, 121–22f
  zero concentration, gas glassy
    polymer systems   115
Transport data
  concept of unrelaxed volume 48–61
  matrix model analysis 116–21
  glassy polymer penetrant   57
  various, discussion 99–101
Transport parameters, CO_2 in lexan
  polycarbonate 65t
Triple point
  CO_2 crystallization 30, 35–41f
  closed-cycle unit crystallization 38, 39f
Tubular membrane element
  configurations   128
Two-column recovery and purification
  process, PSA cell model   195

U

Uptake
  See also Sorption
  complete heat transfer control
    equation   177
  dimensionless, non-isothermal LDF
    model equation 174, 176
  zeolite parametric study, LDF
    model 185–92

V

Vacuum desorption, production of
  high purity nitrogen   156
Vapor phase process, commercial   159
Vapor pressure, various compounds in
  methane recovery 236, 238f
Viscosity
  gas, effect on the separation process   134
  methanol and liquid CO_2 ....35, 36f
Volume
  penetrant effective molecular,
    glassy polymers 53, 54f
  unrelaxed, related to sorption and
    transport properties of glassy
    polymers 48–61

W

Water
  liquid, effect on membrane
    performance 21, 139
  methane solubility at various
    pressures 236, 237t
  removal, CNG AGR process 30, 33
  vapor
    cellulose acetate permeation rates
      and separation factors 134–35t
    removal from hydrocarbon gas
      streams   139

Z

Zeolite
  See also Clinoptilolite
  effective molar or partial molar
    volume of CO_2 ....   55t
  isomer separations 159–63f
  methane and nitrogen separation .213–33
  PSA process for nitrogen
    production 156–60f
  sorption kinetics experimental
    data 178–92
Zinc, regeneration product in
  H_2S removal 360t
Zinc ferrite, sorbent for H_2S removal ..   260
Zinc oxide, sorbent for H_2S removal
  from coal gasification products ..   256

Production by Paula Bérard
Index by Susan Robinson and Anne Riesberg
Jacket design by Kathleen Schaner

Elements typeset by Service Composition Co., Baltimore, MD
Printed and bound by Maple Press Co., York, PA